Preparing a Building Service Life Plan for Green Buildings

About the International Code Council

The International Code Council (ICC) is a member-focused association. It is dedicated to developing model codes and standards used in the design, build, and compliance process to construct safe, sustainable, affordable, and resilient structures. Most U.S. communities and many global markets choose the International Codes. ICC Evaluation Service (ICC-ES) is the industry leader in performing technical evaluations for code compliance fostering safe and sustainable design and construction.

Headquarters: 500 New Jersey Avenue NW, 6th Floor, Washington, DC 20001-2070

District Offices: Birmingham, AL; Chicago, IL; Los Angeles, CA

1-888-422-7233; www.iccsafe.org

Preparing a Building Service Life Plan for Green Buildings

Dru Meadows

McGraw Hill Education

New York Chicago San Francisco
Athens London Madrid
Mexico City Milan New Delhi
Singapore Sydney Toronto

McGraw-Hill Education books are available at special quantity discounts to use as premiums and sales promotions or for use in corporate training programs. To contact a representative, please visit the Contact Us page at www.mhprofessional.com.

Preparing a Building Service Life Plan for Green Buildings

1 2 3 4 5 6 7 8 9 0 DOC/DOC 1 2 0 9 8 7 6 5 4

ISBN 978-0-07-183442-1
MHID 0-07-183442-7

This book is printed on acid-free paper.

Sponsoring Editor
Bridget L. Thoreson

Editing Supervisor
Stephen M. Smith

Production Supervisor
Lynn M. Messina

Project Manager
Asheesh Ratra, MPS Limited

Copy Editor
Surendra Shivam, MPS Limited

Proofreader
A. Nayyer Shamsi, MPS Limited

Art Director, Cover
Jeff Weeks

Composition
MPS Limited

About the Author

Dru Meadows, RA, CCS, FCSI, FASTM, principal with the GreenTeam, Inc., is a specifier, author, teacher, environmentalist, and retired architect with more than 25 years of experience in sustainability consulting and green building. She is a fellow of the Construction Specifications Institute and of ASTM International. Ms. Meadows's work has received recognition from the city of Los Angeles, the state of Oklahoma, the White House, and the United Nations. She is a recognized expert in sustainability standards and has contributed to numerous programs, including the International Organization for Standardization (ISO) standards for service life, for which she chaired the U.S. Technical Advisory Group and served as the American Delegate from 1999 to 2005. Ms. Meadows is the author of the *Federal Green Construction Guide for Specifiers,* with approximately 80 model green specification sections, including a section on facility service life requirements added in December 2010.

Contents

Preface

Green building publications generally advance bold, attention-grabbing challenges with sexy solutions. The reader is transported to a deep green utopia in which, with the right effort, everyone will live sustainably ever after.

This is not one of those books.

Service life planning is not sexy. It is not even new. But its use in green building gives it a new twist that is as stunning as it is unexpected.

There is a practical elegance in a green Building Service Life Plan (BSLP). It is applied life-cycle thinking. It doesn't estimate potential environmental consequences. Rather, it outlines a plan of action to help manage impacts. A green BSLP provides guidance on material maintenance (so the service life can be as long as possible) and reuse/recycling options (so the end-of-life impacts can be managed as sustainably as possible).

The concept of a green BSLP is straightforward. Yet, there remains confusion—among service life professionals, environmentalists, and building industry professionals. Some service life professionals resist the new direction, or seek to marginalize it as secondary to economic issues. Economic issues are important, but they are not central to a green BSLP. Environmentalists often fail to distinguish between Life Cycle Assessment (LCA) and applied life-cycle thinking. LCA is a useful evaluation methodology, but doesn't outline a course of action. A green BSLP provides practical information to help you know what to do. And, there lurks a concern in the building industry that a green BSLP might imply some sort of product guarantee or warranty. After all, it talks about product service life. A good, green BSLP does not in any way provide a guarantee or warranty. It is not a promise. It is a plan. It relies upon the user to implement it.

A green BSLP is the new use of an old tool. This book describes that subtle, but powerful, transformation. It identifies the growing adoption of the green BSLP in green building codes and standards.

And, it offers a template that may be used by green building practitioners along with an example of a complete green BSLP. In developing the example, I relied primarily on the related trade organizations for product durability and maintenance information. They are the experts in their fields. Most of them offer a substantial amount of product guidance publicly available online. The text is heavily footnoted accordingly. Hopefully, this will allow the reader to expand his or her research as may be necessary to develop a project-specific green BSLP. And, assist building product manufacturers in their outreach and communication of green service life issues.

—Dru Meadows

Preparing a Building Service Life Plan for Green Buildings

SECTION 1

Introduction

Will a product work the way it is supposed to? Will it last? These are the basic questions that everyone asks when considering a major purchase. The answers are not entirely dependent on the product manufacture. How we use the product will affect its performance and durability. A well-loved, well-maintained automobile can run smoothly for years. Don't change the oil and you may end up on the side of the road, waiting for assistance. Drive it through a flash flood and you can watch it float away. As any auto mechanic will tell you, if you want your car to provide years of good service, proper maintenance is required.

How are you to know what is "proper maintenance"? Usually, there is an owner's manual. It tells you what performance can be expected, in what conditions, with what type of maintenance. The manual is not a guarantee. It provides information and guidance. It is a kind of service life plan. Most major consumer products today come with an owner's manual describing the service life plan. Buildings do too. Increasingly, a Building Service Life Plan (BSLP) is becoming a core management tool, especially for green buildings.

A BSLP can improve both the economic and the environmental impacts of the building. It improves economic impacts by managing the most costly portion of a building's life-cycle costs, the operation and maintenance stage. Operation and maintenance account for 60 to 85 percent of building life-cycle costs.[1] By guiding appropriate and

---

[1]*Sustainable Federal Facilities: A Guide to Integrating Value Engineering, Life-Cycle Costing, and Sustainable Development*, Federal Facilities Council Technical Report No. 142; 2001, National Academy Press, Washington, D.C., 2001 http://www.wbdg.org/ccb/SUSFFC/fedsus.pdf (accessed November 14, 2013).

According to the report, the total cost of facility ownership, which includes all costs an owner will make over the course of the building's service life, are dominated by operation and maintenance costs, which account for 60 to 85 percent of building life-cycle costs. Design and construction expenditures will account for 5 to 10 percent of building life-cycle costs. Land acquisition, conceptual planning, renewal or revitalization, and disposal will account for 5 to 35 percent of building life-cycle costs.

necessary investments over the course of the building's life cycle, a BSLP increases the residual value of the building.

A BSLP is useful for both existing buildings and new constructions. It helps facility managers by indicating anticipated future operations and cyclical maintenance. It may also be extended to provide cost plans for maintenance. For existing buildings, many of the choices have been predetermined; and, the building is already some way through its service life. Therefore, a BSLP for existing buildings focuses on assessing the residual service life of building elements and optimizing future operation and maintenance. For new buildings, a preliminary BSLP developed during the design phase can help inform critical decisions that will affect the economic and environmental life-cycle impacts of the building. An updated BSLP, revised at the end of construction to reflect the specific building elements incorporated, can help to achieve the targeted design life and manage the associated costs and environmental impacts.

The common objective of service life planning is to help a product function as it is intended to function for the amount of time it is intended to be used. Service life planning is a management tool. It aims to optimize a product's use by identifying the requisite operations and maintenance for the duration of the product's expected lifespan.

Traditionally, a BSLP was used as a life-cycle costing methodology. It was viewed primarily as a design tool for product selection, with a full suite of standards and protocols developed accordingly. More recently, it has been utilized to manage environmental impacts. Because a BSLP examines use, operation, and maintenance across the targeted service life of a building, it provides an established approach to planning for and management of anticipated life-cycle impacts associated with product durability and end-of-life.

From an environmental perspective, a BSLP is the antithesis of planned obsolescence. Planned obsolescence is a marketing strategy that intentionally shortens a product's lifespan so that the customer must purchase replacements more frequently. Often, the lifespan is dictated by fashion. Changing styles drive consumers to replace otherwise functional products with the new model. Sometimes, unavailability of replacement parts or incompatibility with newer components serves to make the product obsolete. Planned obsolescence and waste go hand in hand. While acknowledging that a product may have a short lifespan, a BSLP anticipates the eventual removal/replacement and helps to delay the necessity of removal as well as guide the economic and environmental impacts associated with the product's end-of-life. By examining the maintenance, repair and replacement costs for a specified design life, a BSLP tends to encourage interoperability, adaptability, reuse, and recyclability.[2]

A BSLP for green buildings should describe not only repair and maintenance for promoting the durability of major building elements, but also expectations for their removal, reuse, dismantling, and disposal at their end-of-life. This is consistent with BSLP provisions in the model green building codes and standards.

- The International Green Construction Code (IgCC) 2012 includes a BSLP as a project elective. In accordance with the IgCC elective, a BSLP is essentially a "maintenance, repair and replacement schedule... based on the reference service life data" of the major building elements.[3]
- American Society of Heating, Refrigerating and Air-Conditioning Engineers (ASHRAE) Standard 189.1 delineates requirements for a BSLP. ASHRAE 189.1 Section 10.3.2.3 specifically limits the intent of the BSLP to estimating what components will need to "be repaired or replaced during the service life of the building."
- Green Building Initiative (GBI) Green Globes for New Construction includes a BSLP Credit in the Materials & Resources section of the standard. To achieve the credit, users must identify the reference service life of the building and major building elements. Additionally, they must develop a schedule for maintenance, repair, and replacement for each building element for the duration of the building's design life.[4]

The primary concepts associated with a BSLP are *design life, service life*, and *reference service life*. These have formal, standardized definitions which translate to the vernacular as follows:

- *Design life:* How long you *want* it to last.
- *Service life:* How long it really *does* last.
- *Reference service life:* How long it *should* last (if properly operated and maintained).

Since you can't know the service life until after the fact, a BSLP is written around the design life and the reference service life.

---

[2]For additional information on the standardized application of a BSLP, refer to the International Organization for Standardization (ISO) Standard 15686-6:2004, Buildings and constructed assets—Service life planning—Part 6: Procedures for considering environmental impacts.

[3]Refer to IgCC 2012 Appendix A, paragraphs A105.4 and A105.4.1.

[4]Green Globes for New Construction is based upon the content of the ANSI/GBI standards document, *ANSI/GBI 01-2010: Green Building Assessment Protocol for Commercial Buildings.*

A BSLP does not imply or provide a warranty or guarantee of any kind. It does not ensure the performance of the building or any portion, component, and/or system thereof. It does not guarantee the life-cycle impacts. It is a plan. Like the owner's manual for your car, it provides information and guidance to help you keep your building running smoothly throughout its anticipated life. Additionally, the information in a BSLP can address the end-of-life issues. This is the reason green building practitioners have embraced the BSLP.

BSLPs are attractive for green buildings because they offer a method to extend resource management, especially waste management, across the entire life of a building. The BSLP doesn't require that each product incorporated into a building have an expected lifespan equal to the building's lifespan. It does mean that there should be a plan to recycle or reuse a product when it is removed and/or replaced. A product might be removed in 5 years, 10 years, or 50 years. It might be removed for a variety of reasons. Perhaps it is no longer useful. Perhaps a newer model offers better service. The BSLP can provide guidance on eco-efficient options in removing and reclaiming the product. In this respect, BSLPs address a significant and otherwise unmanaged facet of green building. The BSLP helps to implement life-cycle thinking. It helps close-the-loop through planned management of the end-of-life impacts associated with different materials and components.

The purpose of a BSLP for a green building is to provide a plan for managing the performance requirements and life-cycle impacts (including end-of-life) of the major building elements for the duration of the design life (the intended lifespan) of the building.

A BSLP for a green building should:

- Identify the design life of the building. *Design life* is the intended service life—the period of time targeted for a building or its component parts to meet or exceed the performance requirements.
- Identify the major building elements. Major building elements include items that represent a significant percentage of initial construction cost and/or facility operation cost. Major building elements typically do not include accessory items or equipment and systems related to occupant operations such as furniture, artwork, computers, and manufacturing equipment. Major building elements should include
    - Structural elements
    - Concealed or inaccessible elements
    - Elements for which replacement is cost prohibitive or impractical

    - Major replaceable elements
    - Mechanical, electrical, plumbing, and systems
    - Roofing
    - Site hardscape
- State the reference service life and basis of determination for each major building element. The *reference service life* is the service life that a building or its elements is expected to have under normal in-use conditions. It is important to recognize that a building element need not last as long as the design life of the building. However, the life-cycle impacts, such as anticipated future repairs, removal, reuse, dismantling, and recycling/disposal, should be identified so that they can be managed in accordance with the design intent.
- Describe in-use conditions for each major building element. Products are designed to function in specific applications. These applications are the "in-use conditions" and can dramatically affect the performance and durability of the product. Imagine trying to use bicycle tires on an automobile. Or, snow tires in the tropics. While both types of tires serve the general purpose of a wheel, in order to function optimally, each must be used in the manner intended.
- Describe normal maintenance, repair, and replacement for each major building element for the duration of the building design life. It is never possible to envision every potential problem that might be faced. It is possible to anticipate the maintenance that will be needed due to normal wear and tear.
- Describe the adaptability of each major building element, including the potential for repurposing, reuse, and recycling. What happens at the product's end-of-life? How do we create the infrastructure to cycle our material resources continuously? How do we move from waste minimization to waste elimination? The goal of zero waste is one of the biggest challenges that green building faces.

  "The building industry commandeers 3 billion tons of raw materials annually—40 percent of total global use. It uses almost half of all the mined, harvested, and dredged raw materials each year! Approximately 20 percent of those materials end up as construction waste. That is an estimated 600 million tons of material annually. About a third of that is in the United States alone. The majority of building industry waste is 'clean' waste, scrap materials that have not been used

> but have merely paused momentarily on their brief and fairly direct path from manufacturer to landfill. Think about it. Six hundred million tons of material. That's roughly 400 million cars. Parked end-to-end, that's enough cars to circle the Equator over 60 times. Per year."[5]
>
> A BSLP provides a critical first step in managing the building industry's wastestream long term.

Section Two provides an example of a BSLP for a green building project. The example includes general project information and guidance. Additionally, it includes common major building elements with information organized according to a fill-in-the-blank template as shown below.

BSLP—Major Building Elements Template

Name/Description: ________________

Classification

UNIFORMAT II[6] (*list applicable alpha-numeric designations*) ____________

MasterFormat[7] (*list applicable specification sections*) ________________

Reference Service Life

Years: ________________

Reference Service Life is equal to or greater than Building Design Life

Yes _____ No _____

[5]Meadows, Bell and Whitaker Jr.; RETHINKING RECYCLING: An Economic Perspective of Our Next 109 Years for the Building Industry & Others; 2012, theGreenTeam, Inc.; p. 13.

[6]ASTM E1557-09 Standard Classification for Building Elements and Related Sitework-UNIFORMAT II. ASTM E1557 delineates an organizational framework, called UNIFORMAT II, which is commonly used to organize project information for estimating, planning, and construction management. The classification incorporates hierarchical levels with increasing detail; level 1 is the most general and level 4 the most specific.

[7]MasterFormat-2004 is a standard for organizing construction specifications and other written information for building projects in North America. It is a publication of the Construction Specifications Institute (CSI) and Construction Specifications Canada (CSC). Information is organized by Divisions, with each Division containing more detailed Sections. MasterFormat numbers are 6 digits; the first two represent the division number, the second two represent the general section number, and the last two may be used to further distinguish grouping of information within a general section.

Reference Service Life is equal to or greater than Building Element Design Life

Yes _____ No _____

Basis of determination (*check all that apply*)

_____ Manufacturer's product data describing durability.

_____ Industry data or report(s) for durability compiled by government agency, academic institution, or a related professional or trade organization.

_____ Reference service life established in accordance with ISO 15686 Part 2.

_____ Other information, including professional experience or observation of similar construction or materials in similar conditions, as identified by Architect/Engineer.

Describe: ______________________________

____

____

In-Use Conditions and Other Assumptions

Installation requirements. Installation of building element was in accordance with manufacturer's published instructions and project specifications.

Operation and use. Operation and use is normal for building element.

a. Normal operation and use consists of ______________________________

b. Limitations and restrictions to use for project application are _______

____

c. Maintenance and repair are provided in accordance with the manufacturer's published recommendations and the BSLP maintenance, repair and replacement schedule. Special maintenance access requirements are: ______________________________

Maintenance, Repair, and Replacement Schedule

a. Routine maintenance. Include maintenance instructions, recommended cleaning and accessory materials, and recommended maintenance frequency. ______________________________

____

____

b. Repair and replacement. Identify manufacturer support services for repair and replacement. These may include: diagnostic services, parts availability, upgrade notification procedures, customer service contact information, and online technical support. ____________________

____

____

Adaptability, including limitations on and the potential for

a. Deconstruction and repurposing or reuse during the product design life.

____

____

____

b. Deconstruction and repurposing or recycling at end-of-life. ____________________

____

____

SECTION 2

Building Service Life Plan—Example

This Building Service Life Plan (BSLP) example was developed for a hypothetical green office building in order to include a range of materials typically used in commercial construction. It is divided into four parts.

Part 1: Building Identification. This part includes the basic project data that is commonly found at the front of a project specifications manual and similar documents. It also identifies the design life values for the building and the major building elements.

Part 1 is organized by

- Contact Information
- Project Description
- Building and Major Element Design Life Values

Part 2: BSLP—General. This part describes the scope of the document and the adaptability of the BSLP itself to accommodate future changes or developments that may affect the performance requirements and life-cycle impacts of the major building elements. It includes the whole building statements applicable to operation, maintenance, inspections, repairs, and documentation. And, it identifies general resources for recycling and reuse.

Part 2 is organized by

- Management of Performance Requirements
  - Operation and Maintenance
  - Inspections and Assessments
  - Repairs, Upgrades, and Replacements
  - Documentation

- Management of Life-Cycle Impacts
  - In-Use Impacts
  - End-of-Use Impacts

Part 3: BSLP—Major Building Elements. Part 3 includes BSLP information for each of the common major building elements listed in Part 1. The information is presented in a standardized format according to the fill-in-the-blank template described in the Introduction.

Part 3 is organized by UNIFORMAT II alphanumeric listings.

Part 4: Appendix. Part 4 is intended for additional information that may be applicable to a project BSLP. In this example, the appendix provides background data used in the assessment of some of the reference service life values assigned to major building elements. Part 4 might include unique physical conditions related to the project, long-range community goals for resource management, partnerships dedicated to long-range resource management, and similar related, background material.

PART 1
Building Identification

Contact Information

Project

Name: Foresight Offices

Address: Anywhere, USA

Owner

Name: Wonderful Client

Address: ______

Phone: ______

Contact: ______

Architect/Engineer

Name: Talented Designer

Address: ______

Phone: ______

Contact: ______

Contractor

Name: Affordable Quality Builders

Address: ______

Phone: ______

Contact: ______

Green Consultant

Name:	theGreenTeam, Inc.
Address:	Tulsa, Oklahoma
Phone:	
Contact:	Dru Meadows

Project Description

The project is new construction of a multistory office building. The building is intended to comply with the IgCC 2012 requirements for a green building and will incorporate high-performance technologies to conserve energy, water, and material resources.

Project Data

Construction type	
Gross floor area	
Basement	
Number of stories	
Occupancy group	

Building and Major Building Element Design Life Values

Major building elements include components commonly associated with core and shell construction.

Major building elements include items that represent a significant percentage of initial construction cost and/or facility operation cost.

Major building elements do not include

- Accessory items such as fasteners, sealants, adhesives, paints, and coatings.
- Small, portable items and equipment such as fire extinguishers, mirrors, and window treatment.
- Equipment and systems related to occupant operations such as furniture, artwork, computers, and manufacturing equipment.

Building Design Life (years)	Major Building Element Design Life (years)						
	Structural	**Concealed or Inaccessible**	**Replacement Is Cost Prohibitive or Impractical**	**Major Replaceable**	**Mechanical, Electrical, Plumbing, and Systems**	**Roofing**	**Site Hardscape**
60	**60**	**60**	**60**	**30**	**20**	**20**	**30**
	A1010 Standard Foundations	A2020 Basement Walls	B2010 Exterior Walls	B2020 Exterior Windows	D1010 Elevators and Lifts	B3010 Roof Coverings	G2030 Pedestrian Paving
	A1030 Slab on Grade		C2010 Stair Construction	B2030 Exterior Doors	D20 Plumbing		G2040 Site Development
	B1010 Floor Construction			C1010 Interior Partitions	D30 HVAC		G3030 Storm Sewer
	B1020 Roof Construction			C1020 Interior Doors	D40 Fire Protection		
				C1030 Fittings	D50 Electrical		
				C3010 Wall Finishes			
				C3020 Floor Finishes			
				C3030 Ceiling Finishes			

PART 2

BSLP—General

The purpose of this BSLP is to provide a plan for building operation to manage the *performance requirements* and the *life-cycle impacts (including end-of-life)* of the major building elements for the duration of the design life of the building.

It does not provide a warranty or guarantee for the performance of the building or any portion, component, and/or system thereof. It does not provide a warranty or guarantee for the life-cycle impacts of the building or any portion, component, and/or system thereof.

This BSLP is based upon current information. Standardized databases and tools available through governmental agencies or trade/professional organizations are cited to the extent feasible. It is anticipated that these are more likely to remain constant for the duration of the building design life than are proprietary sources or individual storefronts (whether bricks and mortar or online). It should be noted that in some cases, the readily available industry data cites extremely conservative figures for durability in comparison to empirical information. While there may be empirical evidence that a material lasts hundreds of years and/or professional experience suggests that a material should last much longer than industry statements, this BSLP prioritizes the conservative industry figures. In prioritizing conservative figures, this BSLP takes an approach to "err on the side of caution" with regard in particular to the management of the life-cycle impacts of the major building elements.

Future changes to in-use conditions, evolution of existing processes and technologies, development of new processes and technologies, and other future changes or developments may affect the performance requirements and life-cycle impacts of the major building elements. This BSLP should be adapted in the future as may be necessary to respond to future changes and developments.

Management of Performance Requirements

Management of performance requirements is primarily a function of preventative maintenance for building elements in normal use conditions. Studies demonstrate that up to 30 percent of the energy used in commercial buildings is wasted because of improper operations.[8]

Operation and Maintenance

Facility operation and maintenance (O&M) shall be provided in accordance with the manufacturers' published recommendations, the facility O&M manual and the facility commissioning manual. O&M should be provided to a level 7 in accordance with ASTM E1670.[9]

Inspections and Assessments

Regular inspections shall be performed to assess the serviceability of major building elements and provide recommendations accordingly. Regular inspections shall be performed to assess the continued appropriateness of maintenance procedures and end-of-life procedures for major building elements and provide recommendations accordingly.

Structural building elements and building envelope elements shall be inspected seasonally and rated for serviceability in accordance with ASTM E1700.[10] ASTM E1700 delineates a visual inspection protocol. Where the rating falls below a level 5 for any scale, obtain additional evaluation (including field testing where necessary) from a qualified professional as appropriate to the particular building element(s).

---

[8]Katipamula, S. *Re-Tuning Commercial Buildings: A Low-Cost Path to Energy Efficiency and Cost Savings*, 2010, Pacific Northwest National Laboratory.

Sullivan, G., Pugh, R., Melendez, A. P., and Hunt, W. D. *Operations & Maintenance Best Practices: A Guide to Achieving Operational Efficiency*. 2010, Federal Energy Management Program, U.S. Department of Energy.

[9]ASTM E1670-95a(2012) Standard Classification for Serviceability of an Office Facility for Management of Operations and Maintenance

ASTM E1670 is part of a suite of standards that provides a means to estimate the level of serviceability of a building or facility for one topic of serviceability, and to compare that level against the level of another building or facility. For each requirement, it identifies attributes associated with varying levels of performance. Scales range from levels 1 to 9. Level 7 represents higher than average performance. Level 9 represents the highest.

[10]ASTM E1700-95(2005) Standard Classification for Serviceability of an Office Facility for Structure and Building Envelope

ASTM E1700 is part of a suite of standards that provides a means to estimate the level of serviceability of a building or facility for one topic of serviceability, and to compare that level against the level of another building or facility. For each requirement, it identifies attributes associated with varying levels of performance. Scales range from levels 1 to 9. Level 7 represents higher than average performance. Level 9 represents the highest.

Plumbing building elements shall be inspected seasonally and rated for serviceability in accordance with ASTM E1671.[11] ASTM E1671 delineates a visual inspection protocol. Where the rating falls below a level 5 for any scale, obtain additional evaluation (including field testing where necessary) from a qualified professional as appropriate to the particular building element(s). Additionally, utility bills shall be reviewed monthly against design intent and prior performance. If any utility bill appears excessive, inspect (1) the operation of the system to determine if operation is in accordance with the design intent, and (2) the system itself to assess potential source(s) of excess usage.

Mechanical and electrical building elements shall be regularly inspected and serviced in accordance with the commissioning manual. Additionally, utility bills shall be reviewed monthly against design intent and prior performance. If any utility bill appears excessive, inspect (1) the operation of the system to determine if operation is in accordance with the design intent, and (2) the system itself to assess potential source(s) of excess usage. Retuning may be helpful. Retuning, a method developed by Pacific Northwest National Laboratory (PNNL), aims to reduce energy waste caused by inefficient operations.[12] Retuning utilizes a systematic process to detect operational inefficiencies and implement corrective actions. Retuning is similar to, but less complex than, retro-commissioning and often may be performed by in-house personnel.

With the seasonal inspection for serviceability, review the effectiveness of the BSLP. An important part of the BSLP is to improve the previous decisions regarding the asset so that subsequent operation, maintenance, repair, and replacement activities will be as eco-efficient as possible. At a minimum, the BSLP review should address:

- Changes to in-use conditions
- Updated technologies
- New technologies
- Other developments that may affect the performance requirements and life-cycle impacts of the major building elements

---

[11]ASTM E1671-95a(2012) Standard Classification for Serviceability of an Office Facility for Cleanliness

ASTM E1671 is part of a suite of standards that provides a means to estimate the level of serviceability of a building or facility for one topic of serviceability, and to compare that level against the level of another building or facility. For each requirement, it identifies attributes associated with varying levels of performance. Scales range from levels 1 to 9. Level 7 represents higher than average performance. Level 9 represents the highest.

[12]http://buildingretuning.pnnl.gov/ (accessed December 10, 2013).

If inspections of the major building elements or review of the BSLP indicate that procedures require modification, develop proposed modifications, including a schedule of implementation, to the BSLP accordingly. Coordinate with the O&M manual and the commissioning manual.

Repairs, Upgrades, and Replacements

Repairs, upgrades, and replacements shall be provided as deemed necessary by inspections.

Documentation

Record documents for the facility BSLP, maintenance, inspection, and assessments shall be maintained at the facility.

Management of Life-Cycle Impacts

Buildings and building products striving for sustainability need to pattern their life cycles on the life cycles found in nature. They need to be adaptable and/or durable. They need to anticipate end-of-use and options for reclamation, repurposing, and recycling. And, most importantly, they need a plan to manage their life-cycle impacts. The best of intentions are meaningless if there is no implementation.

In-Use Impacts

Management of in-use life-cycle impacts is directly linked to management of performance requirements. A building element that is performing in accordance with the design intent is expected to demonstrate life-cycle impacts in accordance with the design intent. A properly functioning HVAC system, for example, should demonstrate a level of energy efficiency per the design intent. Therefore, management of in-use life-cycle impacts depends on management of performance requirements.

End-of-Use Impacts

A critical portion of a BSLP is the plan for materials management at end-of-use of the building components and of the building itself. To support a closed-loop life cycle, the plan for end-of-use should address options for reuse, repurposing, and recycling.[13]

---

[13]In a closed-loop life cycle, you capture and reuse or repurpose waste materials.

Per Meadows, Bell and Whitaker Jr.; RETHINKING RECYCLING: An Economic Perspective of Our Next 109 Years for the Building Industry & Others; 2012 the Green Team, Inc.; p. 42:

"How closed is closed? *Closed* may refer to a closed-loop system within a product line, within a manufacturing facility, within a company, within an industry, within a region, or among and between products, facilities, companies, industries and regions. *Closed-loop* may describe reclamation or reuse in a specific and enclosed process; or, it may refer more broadly to utilization of a waste material as a resource."

Reuse and recycling are not interchangeable strategies because reuse is almost always preferable to recycling. The Scottish Ecological Design Association (SEDA) outlines the "hierarchy of waste minimization" for a building as follows, with number one being the highest and best use[14]:

1. Adaptive reuse of existing building
2. Design for adaptability and longevity of new buildings
3. Reuse of building elements/assemblies
4. Reuse of building components
5. Recycling of materials
6. Reclamation of energy from building elements, components, or materials
7. Landfill

General resources for reuse and recycling of building elements include

- **Building Materials Reuse Association (BMRA):** BMRA is a nonprofit educational and research organization whose mission is to facilitate building deconstruction and the reuse/recycling of recovered building materials. Refer to www.bmra.org (accessed November 30, 2012).
- **Construction & Demolition Recycling Association (CDRA):** CDRA promotes the safe and economically feasible recycling of construction and demolition materials such as: concrete, asphalt, asphalt shingles, gypsum wallboard, wood, and metals. CDRA represents a diverse group of member companies and agencies working in the construction materials recycling industry. Refer to www.cdrecycling.org (accessed January 18, 2014).
- **Construction Waste Management Database:** The Database contains information on companies that haul, collect, and process recyclable debris from building projects. Created in 2002 by the U.S. General Services Administration (GSA) Environmental Strategies and Safety Division, the Database is intended to promote responsible waste disposal. Refer to http://www.wbdg.org/tools/cwm.php (accessed September 5, 2012).

[14]SEDA: Scottish Ecological Design Association (SEDA); Design for Deconstruction; SEDA Design Guides for Scotland: No. 1; http://www.seda.uk.net/dfd/dfd.pdf (accessed September 5, 2012).

- **Habitat for Humanity (HFH):** HFH builds and rehabilitates houses with and for its homeowner partner families. The HFH ReStores accept donations of appliances and building materials with a stated mission of providing an environmentally and socially responsible way to keep good, reusable materials out of landfills. Refer to www.habitat.org/restores (accessed November 9, 2012).
- **Institute of Scrap Recycling Industries (ISRI):** ISRI is a trade association representing manufacturers, processors, brokers and industrial consumers of scrap commodities such as ferrous and nonferrous metals, paper, electronics, rubber, plastics, glass, and textiles. Refer to http://www.isri.org (accessed November 26, 2013).
- **Materials Exchanges:** There are international, national, and state/local material exchanges that facilitate the buying and selling of reusable and recyclable materials. Some are for-profit ventures; some are operated by nonprofits or governmental agencies. The U.S. Environmental Protection Agency (EPA) posts a searchable database of material exchanges online. http://www.epa.gov/waste/conserve/tools/exchange.htm (accessed November 25, 2013).
- **National Recycling Coalition (NRC):** NRC is a professional membership and advocacy organization. NRC offers access to state, regional, and national networks. As of October 2012, NRC has affiliation agreements in place with 21 state recycling organizations. Refer to www.nrc-recycle.org (accessed November 7, 2012).
- **Reuse Development Organization (ReDO):** ReDO is a nonprofit organization that promotes reuse of discarded and surplus materials, including building materials. ReDO has an online reuse center locator. Refer to www.redo.org (accessed November 25, 2013).

Additionally, contact the product manufacturer or related trade/professional organization. Most of the building trades either have created or are in the process of creating the infrastructure necessary to support reclamation and recycling in their industry. It is anticipated that over the service life of the building, new reclamation, reuse and recycling infrastructure will be established in the various trades as well as in the building industry as a whole.

PART 3

BSLP—Major Building Elements

A1010 Standard Foundations

Name/Description

Continuous Footings, Spread Footings, Grade Beams, Foundations Walls, and Piers

Classification

UNIFORMAT II—

A1010 Standard Foundations

MasterFormat—

03 20 00 Concrete Reinforcement

03 30 00 Cast-in-Place Concrete

03 30 01 Concrete Accessories

31 63 29 Drilled Piers

Reference Service Life

Years: 66

Reference Service Life is equal to or greater than Building Design Life

Yes ✓ No ___

Reference Service Life is equal to or greater than Building Element Design Life

Yes ✓ No ___

Basis of determination:

_____ Manufacturer's product data describing durability.

__✓__ Industry data or report(s) for durability compiled by government agency, academic institution, or a related professional or trade organization.

_____ Reference service life established in accordance with ISO 15686 Part 2.

__✓__ Other information, including professional experience or observation of similar construction or materials in similar conditions, as identified by Architect/Engineer.

Describe:

Concrete is considered to be extremely durable. Empirical evidence and industry data indicate that a reference service life of 100+ years for cast-in-place concrete is valid. Concrete durability has been defined by the American Concrete Institute (ACI) as its resistance to weathering action, chemical attack, abrasion, and other degradation processes.[15] The ancient Romans used a material that is remarkably close to modern concrete in many of their architectural marvels, such as the Coliseum and the Pantheon. Modern concrete emerged in the late 1800s with the parallel development of reinforced concrete frame construction by G. A. Wayss in Germany, Ernest L. Ransome in the United States, and Francois Hennebique in France. Numerous prominent cast-in-place concrete buildings built in the early to mid-1900s remain functional today.

Most service life modeling for concrete has been developed for use in transportation infrastructure and marine environments. One of the most prominent, consensus-based models is Life-365.

In a 1998 National Institute of Standards and Technology (NIST) workshop, the need for a consensus corrosion service life model was outlined. Subsequently, a consortium was formed within ACI's Strategic Development Council to fund development of the "Life-365" software program. The first version was released in 2001. A second consortium was formed in 2005 to continue development and improvement. The members of that consortium

[15]Guide to Durable Concrete (ACI 201.2R-08).

are the Slag Cement Association (SCA), the Concrete Corrosion Inhibitors Association (CCIA), National Ready Mixed Concrete Association (NRMCA) and the Silica Fume Association (SFA).

Life-365[16] is a standard model developed for predicting the service life and life-cycle costs of reinforced concrete that is exposed to chlorides. The construction types available in the software are limited to transportation structures such as bridges, roadways, marine construction, and parking garages. As in-use conditions are more severe in transportation structures than in typical building structures, it is assumed that the parking garage model will generate a conservative value for the reference service life on this project. A conservative determination assesses the reference service life for concrete as being 66 years.

Life365 v2.0 - Concrete Mixes and Service Lives

Project: SSQ

Description: Default settings for a new project

Analyst: Analyst

Date: 09/25/2012

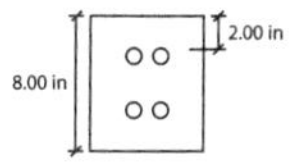

slabs and walls (1-D)

Outer dim: 8 in; clear cover: 2 in

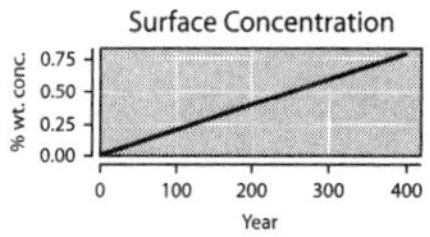

Max surface: 0.8

Years to buildup: 400

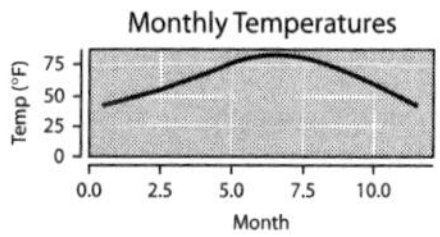

Loc: Texas: DALLAS-FORT WORTH

Type: Parking garages

Concrete Mixes

Alt name	User?	w/cm	SCMs	Inhib.	Barrier	Reinf.
Base case		0.42				Black Steel
Alternative 1		0.42				Black Steel

"n/a" indicates that, since the user is specifying the diffusion properties of this mix, this value is not specified.

Diffusion Properties and Service Lives

Alt name	D28	m	Ct	Init.	Prop.	Service life
Base case	1.38E-8 in*in/sec	0.2	0.05 % wt. conc.	60.2 yrs	6 yrs	**66.2 yrs**
Alternative 1	1.38E-8 in*in/sec	0.2	0.05 % wt. conc.	60.2 yrs	6 yrs	**66.2 yrs**

"->" indicates that the user has directly specified this value; "+" indicates the service life exceeds the study period.

[16]Life-365 is available to download for free.
This example report was developed using Version 2.0. Version 2.2 of the standard model was released July 2013. The Consortium invites industry input for further development of the model. The goals of the development process will be in harmony with ACI Committee 365—Service Life Prediction. Refer to http://www.life-365.org/ (accessed November 14, 2013).

In-Use Conditions and Other Assumptions

Installation requirements. Installation of building element was in accordance with manufacturer's published instructions and project specifications.

Operation and use. Operation and use is normal for building element.

a. Normal operation and use consists of below grade structure.
b. Limitations and restrictions to use for project application are n/a.
c. Maintenance and repair are provided in accordance with the manufacturer's published recommendations and the BSLP maintenance, repair and replacement schedule. Special maintenance access requirements are n/a.

Maintenance, Repair, and Replacement Schedule

a. Routine maintenance. Include maintenance instructions, recommended cleaning and accessory materials, and recommended maintenance frequency.

   ACI states that "Properly designed, proportioned, placed, finished, tested, inspected, and cured concrete is capable of providing decades of service with little or no maintenance."[17]

b. Repair and replacement. Identify manufacturer support services for repair and replacement. These may include diagnostic services, parts availability, upgrade notification procedures, customer service contact information, and online technical support.

   Below grade structure should not require repair or replacement for the duration of building design life. Below grade structure will not be visible for periodic inspections. Where inspections of above grade structure indicate that below grade structure may require repair or replacement, engage a licensed structural engineer to provide evaluation and recommendations.

   Resources for concrete maintenance, repair and replacement include

   AMERICAN CONCRETE INSTITUTE

   38800 Country Club Dr / PO Box 9094

   Farmington Hills, MI 48331

   (248) 848-3700

[17]Guide to Durable Concrete (ACI 201.2R-08).

www.concrete.org

INTERNATIONAL CONCRETE REPAIR INSTITUTE

10600 W. Higgins Rd., Suite 607

Rosemont, IL 60018-3720

(847) 827-0830

www.icri.org

Adaptability, including limitations on and the potential for

a. Deconstruction and repurposing or reuse during the product design life.

   Repurposing and reuse of individual cast-in-place concrete structural elements is not deemed a viable option. However, repurposing and reuse of the whole building may be possible. Engage a licensed structural engineer to assess viability of repurposing and reusing cast-in-place structure for proposed building renovation(s) during the product design life.

b. Deconstruction and repurposing or recycling at end-of-life.

   Repurposing and reuse of individual cast-in-place concrete structural elements is not deemed a viable option. However, recycling and reuse of concrete aggregate is viable. Recycled concrete aggregate has been used in/as aggregate base course (road base), ready mix concrete, asphalt pavement, soil stabilization, pipe bedding, and landscape materials.

   Aggregate base course (road base) is a major market in the United States and an accepted use of recycled concrete by most State Departments of Transportation. ASTM International and American Association of State Highway and Transportation Officials (AASHTO) standards allow use of recycled aggregate. ASTM and AASHTO standards are commonly accepted by agencies across the United States. However, regulatory issues vary from state to state and agency to agency. For example, California wrote legislation mandating the use of recycled aggregate in new concrete; and, New Jersey requires concrete recyclers to obtain a Class B license in order to crush.

Ready mix concrete options are currently limited, but may offer a higher value reuse option in the future as the market evolves. It is anticipated that recyclers will develop eco-efficient methodologies for producing consistent, quality aggregate. Subsequently, ready mix producers will be able to rely better upon the availability of quality material for their end products.

Landscape materials options for recycled concrete aggregate include use of recycled concrete as boulder/stacked rock walls, underpass abutment structures, erosion structures, water features, retaining walls, and decorative hardscape.

Deconstruct to produce as clean a rubble as possible. Depending on the targeted reuse market, rubble may require crushing, pre-sizing, sorting, screening, and contaminant elimination.

Resources for concrete deconstruction, reuse and recycling include

CONCRETERECYCLING.ORG

Sponsored by the Construction & Demolition Recycling Association[18]

1585 Beverly Court, Ste. 112

Aurora, Illinois 60502

(630) 585-7530

www.concreterecycling.org

[18]The Construction & Demolition Recycling Association (CDRA) was founded in 1996 as the Construction Materials Recycling Association (CMRA).

A1030 Slab on Grade

Name/Description

Concrete slab on grade

Classification

UNIFORMAT II—

A1030 Slab on Grade

MasterFormat—

03 20 00 Concrete Reinforcement

03 30 00 Cast-in-Place Concrete

03 30 01 Concrete Accessories

Reference Service Life

Years: 66

Reference Service Life is equal to or greater than Building Design Life

Yes ✓ No ____

Reference Service Life is equal to or greater than Building Element Design Life

Yes ✓ No ____

Basis of determination:

____ Manufacturer's product data describing durability.

✓ Industry data or report(s) for durability compiled by government agency, academic institution, or a related professional or trade organization.

____ Reference service life established in accordance with ISO 15686 Part 2.

✓ Other information, including professional experience or observation of similar construction or materials in similar conditions, as identified by Architect/Engineer.

Describe: See A1010 Standard Foundations

In-Use Conditions and Other Assumptions

Installation requirements. Installation of building element was in accordance with manufacturer's published instructions and project specifications.

Operation and use. Operation and use is normal for building element.

a. Normal operation and use consists of structural use; concrete slab on grade is not intended to function as finished flooring. It will be a substrate for various floor finish materials.

b. Limitations and restrictions to use for project application are install and cure in accordance with project specifications to achieve moisture content compatible with various floor finish materials.

c. Maintenance and repair are provided in accordance with the manufacturer's published recommendations and the BSLP maintenance, repair and replacement schedule. Special maintenance access requirements are n/a.

Maintenance, Repair, and Replacement Schedule

a. Routine maintenance. Include maintenance instructions, recommended cleaning and accessory materials, and recommended maintenance frequency.

ACI states that "Properly designed, proportioned, placed, finished, tested, inspected, and cured concrete is capable of providing decades of service with little or no maintenance."[19]

b. Repair and replacement. Identify manufacturer support services for repair and replacement. These may include diagnostic services, parts availability, upgrade notification procedures, customer service contact information, and online technical support.

See A1010 Standard Foundations

Concrete slab on grade should not require repair or replacement for the duration of building design life. Concrete slab on grade structure will not be visible for periodic inspections. Where inspections of finish flooring applied over slab on grade indicate that slab may require repair or replacement, engage a licensed structural engineer to provide evaluation and recommendations.

[19]Guide to Durable Concrete (ACI 201.2R-08).

Adaptability, including limitations on and the potential for

a. Deconstruction and repurposing or reuse during the product design life.

See A1010 Standard Foundations

b. Deconstruction and repurposing or recycling at end-of-life.

See A1010 Standard Foundations

A2020 Basement Walls

Name/Description

Concrete Basement Walls

Classification

UNIFORMAT II—

A2020 Basement Walls

MasterFormat—

03 20 00 Concrete Reinforcement

03 30 00 Cast-in-Place Concrete

03 30 01 Concrete Accessories

Reference Service Life

Years: 66

Reference Service Life is equal to or greater than Building Design Life

Yes ✓ No ____

Reference Service Life is equal to or greater than Building Element Design Life

Yes ✓ No ____

Basis of determination:

____ Manufacturer's product data describing durability.

✓ Industry data or report(s) for durability compiled by government agency, academic institution, or a related professional or trade organization.

____ Reference service life established in accordance with ISO 15686 Part 2.

✓ Other information, including professional experience or observation of similar construction or materials in similar conditions, as identified by Architect/Engineer.

Describe: See A1010 Standard Foundations

In-Use Conditions and Other Assumptions

Installation requirements. Installation of building element was in accordance with manufacturer's published instructions and project specifications.

Operation and use. Operation and use is normal for building element.

a. Normal operation and use consists of below grade structure.
b. Limitations and restrictions to use for project application are n/a.
c. Maintenance and repair are provided in accordance with the manufacturer's published recommendations and the BSLP maintenance, repair and replacement schedule. Special maintenance access requirements are n/a.

Maintenance, Repair, and Replacement Schedule

a. Routine maintenance. Include maintenance instructions, recommended cleaning and accessory materials, and recommended maintenance frequency.

   ACI states that "Properly designed, proportioned, placed, finished, tested, inspected, and cured concrete is capable of providing decades of service with little or no maintenance."[20]

b. Repair and replacement. Identify manufacturer support services for repair and replacement. These may include diagnostic services, parts availability, upgrade notification procedures, customer service contact information, and online technical support.

   See A1010 Standard Foundations

   Basement walls should not require repair or replacement for duration of the building design life. Basement walls should be visible from the interior of structure for periodic inspections. Where inspections indicate that basement walls may require repair or replacement, engage a licensed structural engineer to provide evaluation and recommendations.

   Concrete deterioration may occur for two principal reasons: corrosion of the embedded steel resulting in concrete deterioration, and degradation of the concrete itself.

   Concrete normally offers protection to embedded reinforcing steel through its alkalinity. Reinforcing steel that is exposed to moisture will corrode over time. The corroded steel expands significantly in volume,

[20]Guide to Durable Concrete (ACI 201.2R-08).

which results in expansive forces on the adjacent concrete, causing it to crack and spall. If deterioration is related to corrosion of the embedded steel, cracking, spalling, and rust staining at the location of the embedded reinforcing steel likely will be evident.

Concrete deterioration due to degradation of the concrete itself is usually related to weathering action, chemical attack, abrasion and similar degradation processes. Exposed concrete is, therefore, more susceptible to such degradation.

Adaptability, including limitations on and the potential for

a. Deconstruction and repurposing or reuse during the product design life.

   See A1010 Standard Foundations

b. Deconstruction and repurposing or recycling at end-of-life.

   See A1010 Standard Foundations

B1010 Floor Construction (Superstructure)

Name/Description

Structural frame, floor decks and slabs, and interior concrete masonry unit (CMU) walls

Classification

UNIFORMAT II—

B1010 Floor Construction (Superstructure)

MasterFormat—

03 20 00 Concrete Reinforcement

03 30 00 Cast-in-Place Concrete

03 30 01 Concrete Accessories

03 38 00 Post-Tension Concrete

04 20 00 Unit Masonry

05 12 00 Structural Steel Framing

05 21 00 Steel Joist Framing

05 21 23 Steel Floor Decking and Shear Studs

05 40 00 Load Bearing Cold-formed Metal Framing

05 50 00 Metal Fabrications

Reference Service Life

Years: 66 years for concrete elements; 100 years for coated steel elements

Reference Service Life is equal to or greater than Building Design Life

Yes ✓ No ____

Reference Service Life is equal to or greater than Building Element Design Life

Yes ✓ No ____

Basis of determination:

____ Manufacturer's product data describing durability.

✓ Industry data or report(s) for durability compiled by government agency, academic institution, or a related professional or trade organization.

_____ Reference service life established in accordance with ISO 15686 Part 2.

__✓__ Other information, including professional experience or observation of similar construction or materials in similar conditions, as identified by Architect/Engineer.

Describe:

For concrete, see A1010 Standard Foundations.

Steel is considered to be extremely durable. Empirical evidence and industry data indicate that a reference service life of 100+ years for coated steel framing is valid. It is not susceptible to rot, termites, or mold. The Steel Framing Institute states that steel used for framing can last from hundreds to over a thousand years due to its zinc coating, a natural element.[21] Recommended coating weights are well defined by industry; refer to AISI S201.[22] A 7-year study conducted by the National Association of Home Builders (NAHB) Research Center measured coating corrosion rates of test samples installed in the exterior wall cavities and ventilated crawl and attic spaces of four houses located in different geographic locations (inland, marine and industrial) in North America. Based on the corrosion rates measured, the estimated life expectancy of the coating varies from 325 years to over 1000 years.[23] A conservative determination assesses the reference service life for coated steel as being 100 years.

CMU is considered to be extremely durable. Empirical evidence and industry data indicate that a reference service life of 100+ years for CMU is valid. CMU construction is regularly used to provide safe-room

[21]Steel Framing Alliance www.steelframing.org/sustainability.html (accessed November 9, 2012).
[22]*North American Standard for Cold-Formed Steel Framing—Product Data*, AISI S201, American Iron and Steel Institute.
[23]ILZRO, ZC-4, *Seven Year Report, Galvanized Steel Framing for Residential Buildings*, 2009.
The project was conducted by the NAHB Research Center under sponsorship of the International Lead Zinc Research Organization, Inc. Research Triangle Park, NC.

construction as it can withstand both routine natural wear as well as extraordinary impacts of natural and human disasters.[24] The National Concrete Masonry Association (NCMA) Tek 8-1A Maintenance of Concrete Masonry Walls (2004) states that the estimated life of CMU is 100+ years.[25] Industry sponsored studies also indicate that a reference service life of 100 years for CMU is valid.[26] CMU consists primarily of portland cement, gravel, sand, and water. An assessment of the reference service life of CMU as equivalent to the reference service life of concrete structure results in a value of 66 years.

In-Use Conditions and Other Assumptions

Installation requirements. Installation of building element was in accordance with manufacturer's published instructions and project specifications.

Operation and use. Operation and use is normal for building element.

a. Normal operation and use consists of above grade structure.
b. Limitations and restrictions to use for project application are n/a.
c. Maintenance and repair are provided in accordance with the manufacturer's published recommendations and the BSLP maintenance, repair and replacement schedule. Special maintenance access requirements are n/a.

Maintenance, Repair, and Replacement Schedule

a. Routine maintenance. Include maintenance instructions, recommended cleaning and accessory materials, and recommended maintenance frequency.

   Concrete: See A1010 Standard Foundations

---

[24]Examples of how concrete masonry can be used in the construction of residential and small community safe rooms are provided in U.S. Federal Emergency Management Agency (FEMA) Publication 320, *Building a Safe Room For Your Home or Small Business; Includes Construction Plans and Cost Estimates,* 3rd ed., August 2008 (see Fig. 5).

[25]The NCMA TEK Series is a compilation of technical topics related to concrete masonry and hardscape design and construction.

[26]PCA R&D Serial No. 2573, Life Cycle Assessment of a Lightweight Concrete Masonry House Compared to a Wood Frame House, by Medgar L. Marceau and Martha G. VanGeem; Portland Cement Association 2002.

CMU: Inspect for the presence of dirt, stains, efflorescence and graffiti and clean as necessary. Care should be taken to use a cleaning method that is as nonaggressive as possible so as not to damage the masonry or surrounding materials. Cleaning by any method should be performed on an inconspicuous section of the building or a sample panel to assess its effect. Provide maintenance in accordance with the National Concrete Masonry Association (NCMA) recommendations:

- 8-1A, Maintenance of Concrete Masonry Walls (2004)
- 8-2A, Removal of Stains from Concrete Masonry (2005)
- 8-3A, Control and Removal of Efflorescence (2003)
- 8-4A, Cleaning Concrete Masonry (2005)

Inspect mortar and units to assess contact between mortar and units. Check for the presence of cracks, chips and other surface degradation. Check plumb and vertical alignment of wall surfaces. If CMU shows evidence of degradation, repair or replacement may be required.

Steel: The Cold-Formed Steel Engineers Institute provides guidance for steel. See information below.[27]

Steel is affected by normal atmospheric elements, especially moisture, which result in the low but continuous oxidization on the steel surfaces, commonly called corrosion. Steel structures in marine atmosphere are more susceptible to corrosion due to the presence of salt in the atmosphere. However, if a building is built to code and properly ventilated and maintained, moisture should not be a concern for galvanized steel framing.

b. Repair and replacement. Identify manufacturer support services for repair and replacement. These may include diagnostic services, parts

[27] Cold-Formed Steel Engineers Institute, TECH NOTE D001-07 September 2007.

availability, upgrade notification procedures, customer service contact information, and online technical support.

Where inspections of the structure indicate that the structural frame, deck, or interior structural CMU walls may require repair or replacement, engage a licensed structural engineer to provide evaluation and recommendations.

Concrete: See A1010 Standard Foundations

CMU: If inspections indicate repair or replacement may be required, refer to NCMA Tek 18-9A Evaluating Existing CM Construction (2003). This TEK outlines industry practices for the physical evaluation of masonry units, grout, mortar, and assemblies in a building.

Steel: If degradation of the coating is evident, provide maintenance of structural steel coatings in accordance with the Society for Protective Coatings (SSPC) SSPC-PA Guide 5 Guide to Maintenance Coating of Steel Structures.[28] This guide covers procedures for developing a maintenance coating program for steel structures. The guide may be used for one-time recoat programs or long-range recoat programs.

Adaptability, including limitations on and the potential for

a. Deconstruction and repurposing or reuse during the product design life.

Concrete: See A1010 Standard Foundations

CMU: With care in deconstruction individual units can be reused. Mortared units should be carefully cleaned of old mortar before reuse. Reused units should be tested to verify that they meet the requirements of the current specification for their intended use. Units that will be laid in mortar must have adequate open pores to ensure proper bonding.

[28]SSPC was founded in 1950 as the Steel Structures Painting Council. In 1997, the name of the association was changed to The Society for Protective Coatings to reflect the changing and expanding industry.

Steel: When dismantling steel members for recycling, it is common to cut the steel members down to small segments that are easily transported and sold. However, for steel to be reused or repurposed, it should be recovered in as much of its original intact length as possible.

For reuse in structural applications: Identify the salvage location, documenting the location in the building structure from where the members were salvaged, including the date of construction of original building. Members should not have areas of accelerated corrosion or show evidence of localized section loss. Members should be tested and certified to the section properties and material grade as required for repurposing. Appendix 5, "Evaluation of Existing Structures," of AISC 360-10 provides a starting point for evaluating salvaged structural steel.[29] While geared toward renovation projects, the guidelines in the appendix may be applied to repurposing of salvaged structural steel for new construction.

b. Deconstruction and repurposing or recycling at end-of-life.

Concrete: See A1010 Standard Foundations

CMU: CMU can be crushed and recycled into aggregate for landscaping, as subbase material for pavements, or as aggregate for concrete. CMU recycling is similar to concrete recycling; see A1010 Standard Foundations

Steel: All steel products, including steel framing and steel roofing, contain recycled steel. Steel framing contains a minimum of 25 percent recycled steel and is continually and completely recyclable. Steel is the most recycled material in the world. In 2012, 75 million metric tons of

[29] *2010 Specification for Structural Steel Buildings*, ANSI/AISC 360-10; refer to www.aisc.org/WorkArea/showcontent.aspx?id=26516 (accessed November 7, 2012). This specification provides common requirements for the design and construction of structural steel buildings.

iron/steel scrap was recycled in the United States, representing more than 55 percent of the volume of all domestically processed material.[30]

Resources for steel deconstruction, reuse and recycling include

AMERICAN INSTITUTE OF STEEL CONSTRUCTION

One East Wacker Drive Suite 700

Chicago, IL 60601-1802

312.670.2400

www.aisc.org

THE STEEL RECYCLING INSTITUTE

(800) 876-7274

www.recycle-steel.org

The Steel Recycling Institute operates the Steel Recycling Database which is a searchable online database of recycling locations/organizations across the United States.[31]

---

[30]Institute of Scrap Recycling Industries, The Scrap Recycling Industry: Iron and Steel 2013 Fact Sheet http://www.isri.org/ISRI/_About/Scrap_Recycling_Industry/ISRI/_About/Scrap_Recycling_Industry.aspx?hkey=1c15f9cb-6f70-4130-a053-ceba7a327c7f (accessed November 26, 2013).

[31]www.recycle-steel.org/Recycling%20Resources/Locator.aspx (accessed December 4, 2013).

B1020 Roof Construction (Superstructure)

Name/Description

Roof framing and decking

Classification

UNIFORMAT II—

B1020 Roof Construction (Superstructure)

MasterFormat—

03 20 00 Concrete Reinforcement

03 30 00 Cast-in-Place Concrete

03 30 01 Concrete Accessories

05 12 00 Structural Steel Framing

05 21 00 Steel Joist Framing

05 31 23 Steel Roof Decking

05 40 00 Load Bearing Cold-Formed Metal Framing

05 50 00 Metal Fabrications

Reference Service Life

Years: 66 years for concrete elements; 100 years for coated steel elements.

Reference Service Life is equal to or greater than Building Design Life

Yes ✓ No ____

Reference Service Life is equal to or greater than Building Element Design Life

Yes ✓ No ____

Basis of determination:

____ Manufacturer's product data describing durability.

✓ Industry data or report(s) for durability compiled by government agency, academic institution, or a related professional or trade organization.

____ Reference service life established in accordance with ISO 15686 Part 2.

✓ Other information, including professional experience or observation of similar construction or materials in similar conditions, as identified by Architect/Engineer.

Describe: See B1010 Floor Construction (Superstructure)

In-Use Conditions and Other Assumptions

Installation requirements. Installation of building element was in accordance with manufacturer's published instructions and project specifications.

Operation and use. Operation and use is normal for building element.

a. Normal operation and use consists of above grade structure.
b. Limitations and restrictions to use for project application are n/a.
c. Maintenance and repair are provided in accordance with the manufacturer's published recommendations and the BSLP maintenance, repair and replacement schedule. Special maintenance access requirements are n/a.

Maintenance, Repair, and Replacement Schedule

a. Routine maintenance. Include maintenance instructions, recommended cleaning and accessory materials, and recommended maintenance frequency.

   See B1010 Floor Construction (Superstructure)

b. Repair and replacement. Identify manufacturer support services for repair and replacement. These may include diagnostic services, parts availability, upgrade notification procedures, customer service contact information, and online technical support.

   See B1010 Floor Construction (Superstructure)

Adaptability, including limitations on and the potential for

a. Deconstruction and repurposing or reuse during the product design life.

   See B1010 Floor Construction (Superstructure)

b. Deconstruction and repurposing or recycling at end-of-life.

   See B1010 Floor Construction (Superstructure)

B2010 Exterior Walls

Name/Description

Exterior Walls—Masonry, stone, and plaster

Classification

UNIFORMAT II—

B2010 Exterior Walls

MasterFormat—

04 20 00 Unit Masonry

04 42 00 Exterior Stone Cladding

04 72 00 Cast Stone Masonry

09 24 00 Cement Plastering

09 75 13 Stone Paneling

09 75 19 Stone Trim

Reference Service Life

Years: 100 for clay masonry, 100 for cast stone masonry, 100 for natural stone, 66 for cement plaster

Reference Service Life is equal to or greater than Building Design Life

Yes ✓ No ____

Reference Service Life is equal to or greater than Building Element Design Life

Yes ✓ No ____

Basis of determination:

____ Manufacturer's product data describing durability.

✓ Industry data or report(s) for durability compiled by government agency, academic institution, or a related professional or trade organization.

____ Reference service life established in accordance with ISO 15686 Part 2.

✓ Other information, including professional experience or observation of similar construction or materials in similar conditions, as identified by Architect/Engineer.

Describe:

Brick masonry: Brick masonry is considered to be extremely durable. Empirical evidence and industry data indicate that a reference service life of 100+ years for brick masonry is valid. Dried clay brick construction dating as early as 7500 to 7000 BC has been discovered in the Middle East. Fired clay brick construction dates back to 4500 BC in the Middle East. Fired clay brick was used by the ancient Greeks and Romans. Many examples of brick construction dating back hundreds of years remain functional today. According to the Brick Industry Association, clay brick masonry "is extremely durable, having a life expectancy of hundreds of years."[32] A conservative determination assesses the reference service life for clay masonry as being 100 years.

Cast stone masonry: The Cast Stone Institute provides guidance for cast stone. See information below.[33]

Cast stone is an architectural building product that combines the strength and durability of reinforced precast concrete with the appearance of natural stone. It consists of portland cement, fine and coarse aggregates (usually of granite, quartz or limestone), sand and high-performance chemical admixtures. Cast stone was used as early as 1138 in Carcassonne, France, the Cité. In 1900, its use became common in London; and, by 1920, in the United States. The industry has established performance standards for durability and resistance to freeze-thaw, water absorption, abrasion, and dirt. According to the Cast Stone Institute, cast stone compliant with industry standards has a service life exceeding 100 years.

[32]BIA, Technical Notes on Brick Construction #48, June 2009.

[33]Cast Stone Institute; Basic Comparisons between Various Masonry Materials Design Tips, Technical Bulletin #49.

Stone: Natural stone is considered to be extremely durable. Empirical evidence and industry data indicate that a reference service life of 100+ years for stone is valid. The Natural Stone Council has developed life-cycle datasets for granite, limestone, sandstone and slate quarrying, and processing operations.[34] Using the life-cycle data, the Council has produced material fact sheets describing several stone types. Both granite and limestone are reported to last 100 years or more. A conservative determination assesses the reference service life for stone as being 100 years.

Cement plaster: Portland cement plaster is a combination of portland cement-based cementitious material, sand, and water. For centuries, it has been used as a coating for walls and ceilings and for decoration. It is considered to have the "toughness and durability of concrete."[35] Practitioners estimate that it has a typical service life of 50 to 80 years or more. An assessment of the reference service life of portland cement plaster as equivalent to the reference service life of concrete structure results in a value of 66 years.

In-Use Conditions and Other Assumptions

Installation requirements. Installation of building element was in accordance with manufacturer's published instructions and project specifications.

Operation and use. Operation and use is normal for building element.

a. Normal operation and use consists of exterior envelope.
b. Limitations and restrictions to use for project application are n/a.
c. Maintenance and repair are provided in accordance with the manufacturer's published recommendations and the BSLP maintenance, repair and replacement schedule. Special maintenance access requirements are n/a.

[34]The datasets can be downloaded from the University of Tennessee Center for Clean Products. Refer to: http://isse.utk.edu/ccp/projects/naturalstone/results_pubs.html (accessed November 18, 2013).

[35]Portland Cement Association; http://www.cement.org/homes/ch_bs_stucco.asp (accessed November 13, 2012).

Maintenance, Repair, and Replacement Schedule

a. Routine maintenance. Include maintenance instructions, recommended cleaning and accessory materials, and recommended maintenance frequency.

Brick masonry: Provide routine maintenance in accordance with the Brick Industry Association recommendations.[36] See BIA, Technical Notes on Brick Construction:

- #20, Cleaning Brickwork
- #23, Stains—Identification and Prevention
- #23A, Efflorescence—Causes and Prevention
- #46, Maintenance of Brick Masonry

Cast stone masonry: Provide routine maintenance in accordance with the Cast Stone Institute recommendations.[37] See technical bulletins:

- #32, Crazing
- #33, Efflorescence
- #36, Inspection and Acceptance
- #39, Cleaning

Stone: Routine cleaning should be in accordance with ASTM C1515.[38] This guide describes materials and procedures which may be used to clean dimension stone exteriors. It relates primarily to routine maintenance of the surface finish and is not intended for use in the restoration of disfigured, severely soiled or stained stonework, or of stonework of particular historic importance.

[36]BIA, Technical Notes on Brick Construction; www.gobrick.com (accessed November 13, 2012).
[37]Cast Stone Institute; www.caststone.org (accessed November 13, 2012).
[38]ASTM C1515-11 Standard Guide for Cleaning of Exterior Dimension Stone, Vertical And Horizontal Surfaces, New or Existing.

Cement plaster: The Stucco Manufacturers Association (SMA) provides guidance for cement plaster. See information below.[39]

SMA outlines a four-step process for routine cleaning of exterior plaster walls:

1. Pre-wet the wall, saturating it. Start at the bottom and work to the top.
2. Pressure wash with water to loosen the dirt. Start at the top and wash the dirt down the wall. Caked on dirt may require brushing.
3. Mild, water soluble cleaners may be used to remove stains.
4. Rinse remaining dirt off the wall.

SMA provides additional guidance for cleaning efflorescence as follows. Most efflorescence can be removed by dry brushing and rinsing with clean water. However, efflorescence caused by calcium carbonate, which appears usually as a white "bloom" or crust, is more difficult to remove. Removal of this type of efflorescence may require saturation with water followed by a wash with a diluted acid, followed immediately with an alkaline wash, and then a clean water rinse. Removal of calcium carbonate efflorescence should be provided by a trained professional.

b. Repair and replacement. Identify manufacturer support services for repair and replacement. These may include diagnostic services, parts availability, upgrade notification procedures, customer service contact information, and online technical support.

Brick masonry: The Brick Industry Association (BIA) provides guidance on brick masonry. See information below.[40]

---

[39]Stucco Manufacturers Association (SMA), http://stuccomfgassoc.com/ (accessed November 18, 2013). SMA technical papers include

- Judging Stucco Installations
- Efflorescence
- Three-Coat Stucco Maintenance Guide
- Stucco Crack Policy
- Care & Maintenance of Stucco

[40]BIA, Technical Notes on Brick; www.gobrick.com (accessed November 13, 2012).

BIA, Technical Notes On Brick Construction #46, Maintenance of Brick Masonry covers maintenance of brick masonry with an emphasis on preventing moisture penetration. Properly designed and installed brick masonry should require little maintenance. Components incorporated in the brickwork (such as caps, copings, sills, lintels and sealant joints) may require periodic repair. Neglecting maintenance of such components may lead to deterioration of the brickwork. Mortar used in walls should last 25+ years; re-pointing may be required every 25 to 50 years or more.[41]

Cast stone masonry: Provide maintenance and repair in accordance with the Cast Stone Institute recommendations.[42] See technical bulletins:

#38, Touchup and Repair

Stone: The Marble Institute of America (MIA) publishes *A Guide to the Care & Cleaning of Natural Stone.*[43] The guide is applicable to maintenance and repair of all types of stone.

The Indiana Limestone Institute (ILI) provides guidance for stone. See information below.[44]

ILI recommends that small chips be left alone, especially if they do not detract from the appearance of the finished work. Larger chips may be repaired using cementitious materials made for the purpose. When the chip can be saved, it is usually possible to replace it using either a thermo-setting resin adhesive, or cyanoacrylate "superglue." Alternately,

[41]BIA, Technical Notes on Brick Construction #46 and #48. Technical Notes #48 includes a checklist for general inspections and specific repair techniques. It also includes a table with estimated repair times for brickwork.

[42]Cast Stone Institute; www.caststone.org (accessed November 13, 2012).

[43]Marble Institute of America (MIA), www.marble-institute.com. MIA is a resource for information and education for the natural stone industry, not just marble.

[44]Indiana Limestone Institute of America, http://www.iliai.com/.

The Institute publishes *Repairing Indiana Limestone.* The Institute Technote Series includes:

- Cleaning Indiana Limestone
- Damage and Repair Practices and Standards
- Durability and Weathering in Contemporary Atmospheres

a "dutchman" can be used to repair the area. A dutchman is a separate piece of stone cut to fit tightly in place of the damaged portion.

Broken corners can usually be repaired using thermo-setting adhesives. In some cases, dowels, plates, or angles also may be required.

Cracked stones may be repaired. However, care should be taken to determine that it remains structurally stable.

Cement plaster: The Portland Cement Association (PCA) provides guidance for cement plaster. See information below.[45]

Moisture management is critical because trapped or excess water can lead to rust and rot. Portland cement plaster is a breathable material that does allow moisture vapor to pass through, so it is capable of drying out. PCA publishes the Portland Cement Plaster/Stucco Manual which includes a troubleshooting guide and relevant industry standards. Where inspection indicates repair may be necessary, detailed evaluation and repair should be provided by a trained professional. Sources for industry professionals include

SOUTH CENTRAL WALL, CEILING & PLASTER ASSOCIATION
320 Westway Place, Ste. 501
Arlington, Texas 76018
(817) 461-0676; Toll Free: (866) 96-TLPCA
info@tlpca.org

Adaptability, including limitations on and the potential for

a. Deconstruction and repurposing or reuse during the product design life.

Brick masonry: With care in deconstruction, brick masonry units can be reused. Mortared units should be carefully cleaned of old mortar

[45]Portland Cement Association (PCA), http://www.cement.org/.

before reuse. Units that will be laid in mortar must have adequate open pores to ensure proper bonding. Reused units should be tested to verify that they meet the requirements of the current specification for their intended use. See BIA, Technical Notes on Brick Construction[46]:

- #15, Salvaged Brick
- #48, Sustainability and Brick

Cast stone masonry: With care in deconstruction, cast stone elements can be reused. Copings, sills, and other shapes can be reused if cleaned of mortar. Cast stone can be repaired to conceal damage that may occur during disassembly and removal.

Stone: With care in deconstruction, stone elements can be reused. Stone building elements have been reused and repurposed for hundreds of years. The reclaimed material may be cut and finished with the same methods and standards as raw material.

Cement plaster: Repurposing and reuse of cement plaster wall finishes is not deemed a viable option.

b. Deconstruction and repurposing or recycling at end-of-life.

Brick masonry: Brick can be crushed and recycled into new brick or used as aggregate for landscaping or as subbase material for pavements.

Cast stone masonry: Cast stone masonry can be crushed and used as aggregate for landscaping or subbase material for pavements.

Stone: Stone can be crushed and recycled for landscaping, as subbase material for pavements, or as aggregate for terrazzo or concrete.

[46]BIA, Technical Notes on Brick Construction; www.gobrick.com (accessed November 13, 2012)

Cement plaster: Repurposing and reuse of cement plaster wall finishes is not deemed a viable option. However, recycling and reuse of cement plaster wall finishes may be viable. Portland cement plaster may applied directly to a solid base such as masonry or concrete walls. It is also commonly applied over a metal lath, water-resistant building paper, and framing. The material components of this plaster wall assembly (cementitious aggregate, metal, and paper) are considered to be recyclable. However, the components are not typically separated during deconstruction but are removed together as mixed waste. Verify availability of recycling facilities for such mixed waste prior to deconstruction.

B2020 Exterior Windows

Name/Description

Exterior windows and glazing systems

Classification

UNIFORMAT II—

B2020 Exterior Windows

MasterFormat—

08 41 26 All-Glass Entrances And Storefronts

08 44 13 Glazed Aluminum Curtain Walls

08 49 50 Folding Glass Wall Systems

08 80 00 Glazing

Reference Service Life

Years: 40

Reference Service Life is equal to or greater than Building Design Life

Yes ____ No ✓

Reference Service Life is equal to or greater than Building Element Design Life

Yes ✓ No ____

Basis of determination:

____ Manufacturer's product data describing durability.

✓ Industry data or report(s) for durability compiled by government agency, academic institution, or a related professional or trade organization.

____ Reference service life established in accordance with ISO 15686 Part 2.

✓ Other information, including professional experience or observation of similar construction or materials in similar conditions, as identified by Architect/Engineer.

Describe:

Component materials commonly used in a curtain wall system are very durable. Aluminum and glass, for example, may have very long life spans. The U.S. Environmental Protection Agency (EPA) states that an aluminum can will not degrade for 200 years and a glass bottle will not degrade for an undetermined [indefinite] period of time.[47] Aluminum frames are inherently corrosion resistant in many environments if anodized and properly sealed or painted with baked-on fluoropolymer paint. Glass, if unbroken, may serve its intended function indefinitely.

The reference service life of curtain wall systems, however, are dependent more on the design of the system than on the durability of the component materials. Industry practitioners assess service life of curtain walls as 30 to 40 years.[48] According to the University of Washington, the reference service life of curtain walls, storefronts and windows is 50 years. This includes all joints within the curtain wall or window. Perimeter joints at the interface with other exterior building systems is 10 years for joint sealants and

---

[47]U.S. EPA http://www.epa.gov/gmpo/edresources/debris_t.html (accessed November 14, 2012).

Sources for rates of decomposition of litter (trash) typically reference one or more of the following:

1. The New York Times (Nemve E. Metropolitan Diary, October 1, 2001): Paper—2.5 months; orange peel—6 months; milk carton—5 years; cigarette butt—10 to 12 years; plastic bag—10 to 20 years; disposable diaper—75 years; tin can—100 years; beer can—200 to 500 years; Styrofoam—never.
2. Penn State University (per the U.S. National Park Service; Mote Marine Lab, FL and "Garbage In, Garbage Out," Audobon Magazine, *Sept/Oct 1998*): Paper—2 to 4 weeks; leaves—1 to 3 months; orange peel—6 months; milk carton—5 years; plastic bag—10 to 20 years; plastic container—50 to 80 years; aluminum can—80 years; tin can—100 years; plastic soda bottle—450 years; glass bottle—500 years; Styrofoam—never.
3. "Pocket Guide to Marine Debris," The Ocean Conservancy, 2004: Paper towel—2 to 4 weeks; orange or banana peel—2 to 5 weeks; newspaper—6 weeks; apple core—2 months; waxed milk carton—3 months; plywood—1 to 3 years; wool sock—1 to 5 years; cigarette filter—1 to 50 years; plastic bag—10 to 20 years; plastic film canister—20 to 30 years; nylon fabric—30 to 40 years; leather—50 years; tin can—50 years; foamed plastic cup—50 years; rubber boat sole—50 to 80 years; foamed plastic buoy—80 years; aluminum can—80 to 200 year; disposable diapers—450 years; plastic beverage bottles—450 year; plastic beverage bottles—450 year; monofilament fishing line—600 years; glass bottle—1,000,000 years.

[48]Stuart Dean, White Paper: Anodized Aluminum Curtain Walls Preventative Maintenance, www.stuartdean.com (accessed November 14, 2012).

50 years for cavity flashing.[49] GreenSpec states that the reference service life is 60+ years for aluminum extrusions, 60+ years for glass (although insulating glass may be less), and 15 to 30+ years for gaskets and sealants.[50]

Conservatively, the reference service life of an aluminum curtain wall system may be estimated to be 40 years with maintenance (and replacement as necessary) of gaskets and sealants.

In-Use Conditions and Other Assumptions

Installation requirements. Installation of building element was in accordance with manufacturer's published instructions and project specifications.

Operation and use. Operation and use is normal for building element.

a. Normal operation and use consists of exterior envelope; and, pedestrian access and egress.
b. Limitations and restrictions to use for project application are n/a.
c. Maintenance and repair are provided in accordance with the manufacturer's published recommendations and the BSLP maintenance, repair and replacement schedule. Special maintenance access requirements are

The curtain wall should be designed to accommodate access for maintenance. Low-rise buildings may be accessed from the ground using equipment with articulated arms. High rise buildings should have swing stage access for window cleaning, and routine maintenance and repair. Davits and fall arrest safety tieback anchors should be provided on the roof, and

[49]University of Washington, Facilities Services Design Guide: Curtainwalls and Windows, https://www.washington.edu/facilities/engr/fsdg/04_Architectural/06-8A_Curtainwalls&Windows.pdf (accessed November 14, 2012).

[50]GreenSpec is a green building product directory operated by BuildingGreen.com. http://www2.buildinggreen.com/frequently-asked-questions (accessed November 14, 2012):

Aluminum extrusions, usually based on alloy 6063 to BS EN 755–1, has a reference service life of more than 60 years. The type of finish dictates life to first maintenance or replacement:

- Anodising to BS 3987, minimum 25 microns thick, 30 years or more.
- Polyester powder coating to BS 6496, minimum 40 microns thick, 15 to 25 years.
- Polyvinylidene–fluoride (PVDF or PVF2) minimum 25 microns thick, 20 years or more.

Glass has a reference service life of 60+ years. However, insulated glass has shorter expectancies due to edge seal failures and infiltration of water vapor into the space between the panes. Insulating glass units to BS EN 1279 should be specified to give an expected service life of 25 years.

Noncellular EPDM gaskets to BS 4255 have a reference service life 15 to 30 years or more. Neutral cure type silicone sealant has a reference service life of 5 to 20 years.

stabilization tie-offs provided on the face of the wall to comply with OSHA standards CFR 1910.66, CFR 1910.28 and ANSI/IWCA I-14.1 "Window Cleaning Safety Standard."

Maintenance, Repair, and Replacement Schedule

a. Routine maintenance. Include maintenance instructions, recommended cleaning and accessory materials, and recommended maintenance frequency.

   Provide maintenance of aluminum frame in accordance with the American Architectural Manufacturers Association (AAMA) recommendations.[51] See AAMA 609 & 610, Cleaning and Maintenance Guide for Architecturally Finished Aluminum.

   The Glass Association of North America (GANA) provides guidance on glazing. See information below.[52]

   Cleaning materials should be mild and nonabrasive. Since glass products can be permanently damaged if improperly cleaned, glass producers, and fabricators recommend procedures for properly cleaning glass surfaces.

   - Proper Procedures for Cleaning Architectural Glass Products—GANA 01-0300
   - Guidelines for Handling and Cleaning Decorative Glass—DD 01-0608
   - Suggested Procedures for Dealing with Broken Glass—GANA 04-0906

b. Repair and replacement. Identify manufacturer support services for repair and replacement. These may include diagnostic services, parts availability, upgrade notification procedures, customer service contact information, and online technical support.

---

[51]American Architectural Manufacturers Association,www.aamanet.org (accessed November 14, 2012).

[52]Glass Association of North America (GANA) http://www.glasswebsite.com/ (accessed November 18, 2013).

The U.S. Whole Building Design Guide (WBDG) provides guidance on curtain walls. See information below.[53]

Factory finished aluminum frames are resistant to environmental degradation and require only periodic cleaning. Recoating factory painted aluminum is possible with an air-dry fluoropolymer coating; but the recoating will not be as durable as the baked-on original coating. Anodized aluminum frames cannot be "re-anodized" in the field. A clear coating may be applied.

Perimeter gaskets and sealants, properly designed and installed, have a reference service life of 10 to 20 years. Properly installed silicone wet seals should last 10 to 20 years; gaskets 15 to 20 years.

Internal gaskets and sealants may fail due to curtain wall movements, moisture penetration, and/or UV degradation. Repair of the internal seal will require disassembly of the curtain wall. Alternately, installation of exterior surface wet sealing at all glazing and frame joints may be an effective solution.

Where inspection indicates repair may be necessary, detailed evaluation and repair should be provided by a trained professional.

Adaptability, including limitations on and the potential for

a. Deconstruction and repurposing or reuse during the product design life. Deconstruction for reuse may be feasible; however, deconstruction for recycling is more likely. See deconstruction at end-of-life.
b. Deconstruction and repurposing or recycling at end-of-life.

---

[53]Whole Building Design Guide (WBDG). The WBDG is a web-based portal that provides information on a wide range of building-related topics from a "whole buildings" perspective. It is organized into three major categories—Design Guidance, Project Management and Operations & Maintenance. www.wbdg.org (accessed November 14, 2012).

Aluminum and steel: The metal components of curtain wall systems (aluminum and steel) are considered to be recyclable. However, the components are not typically separated during deconstruction but are removed together as mixed waste. Verify availability of recycling facilities for such mixed waste prior to deconstruction.

Glazing: The float glass industry does not have an infrastructure (similar to that of the steel and aluminum industry) to reclaim postconsumer glazing. Although glass is a recyclable material, recycling glazing from buildings has a couple of key barriers:

- Abundance of raw materials. The raw materials are abundant and comparatively inexpensive. Silica (sand), for example, comprises 73 percent on average of glazing material.
- Irregular cullet composition in glazing waste. The chemical composition of glass used in buildings is different from that in windshields or beverage containers. Using postconsumer glass waste is problematic for manufacturing since it will include a variety of glass with unknown chemical properties, as well as laminates and coatings. The mixed composition waste can cause furnace downtime and product quality issues.

Reclamation of glazing is gaining momentum. In July 2012, InnoCentive[54] issued a Challenge to solicit ideas on recycling glass from

[54]InnoCentive is a global innovation marketplace that engages the public to solve some of the world's most important problems for cash awards up to $1 million. InnoCentive poses challenges in areas ranging from the environment to medical advancements. An InnoCentive Ideation™ Challenge is a broad question formulated to obtain access to new ideas, similar to a global brainstorm for producing a breakthrough idea or market survey which may include ideas for a new product line, a new commercial application for a current product, or even a viral marketing idea to recruit new customers.

Challenge ID: 9932970: Recycling Challenge: How to Recover Glass from Buildings? https://www.innocentive.com/ar/challenge/9932970 (accessed December 4, 2012).

buildings. Glass for Europe, the trade association for Europe's manufacturers of building, automotive, and transport glass, reports that new and innovative recycling technologies are emerging in response to the EU Directive on end-of-life vehicles (2000/53/EC).[55]

Resources for glazing deconstruction, reuse and recycling include

AMERICAN ARCHITECTURAL MANUFACTURERS ASSOCIATION
1827 Walden Office Square, Suite 550
Schaumburg, Illinois 60173-4268
(847) 303-5664
www.aamanet.org

GLASS ASSOCIATION OF NORTH AMERICA
2945 SW Wanamaker Drive, Suite A
Topeka, KS 66614-5321
(785) 271-0208
www.glasswebsite.com

[55]http://www.glassforeurope.com/images/cont/167_67296_file.pdf (accessed December 4, 2012).

B2030 Exterior Doors

Name/Description

Exterior doors

Classification

UNIFORMAT II—

B2030 Exterior Doors

MasterFormat—

08 11 13 Hollow Metal Doors and Frames

08 11 19 Stainless Steel Doors And Entrances

08 41 13 Aluminum-Framed Entrances and Storefronts

08 71 00 Door Hardware

Reference Service Life

Years: 30

Reference Service Life is equal to or greater than Building Design Life

Yes ____ No ✓

Reference Service Life is equal to or greater than Building Element Design Life

Yes ✓ No ____

Basis of determination:

____ Manufacturer's product data describing durability.

✓ Industry data or report(s) for durability compiled by government agency, academic institution, or a related professional or trade organization.

____ Reference service life established in accordance with ISO 15686 Part 2.

✓ Other information, including professional experience or observation of similar construction or materials in similar conditions, as identified by Architect/Engineer.

Describe:

The Steel Door Institute (SDI) provides guidance for steel doors. See information below.[56]

According to SDI, properly installed and maintained hollow metal doors last 30 to 40 years. Durability is a function of door and frame construction (e.g., size and thickness of aluminum frames, thickness of steel sheet facings) and the durability of door hardware. Corrosion resistance depends upon the coating system. Coating systems that include a galvanic protection primer (e.g., zinc-rich paint, hot dipped galvanizing) in combination with a barrier coat of paint are more durable than those that rely on a barrier coat of paint alone. Hollow steel doors with direct exposure to the weather may have shorter service lives because water tends to penetrate between the exterior and interior steel sheets, causing the doors to rust from the inside out.

Conservatively, the reference service life of exterior hollow metal doors may be estimated to be 30 years with maintenance (and replacement as necessary) of hardware, weatherstripping, gaskets, and sealants.

Stainless Steel: See C1030 Fittings

In-Use Conditions and Other Assumptions

Installation requirements. Installation of building element was in accordance with manufacturer's published instructions and project specifications.

Operation and use. Operation and use is normal for building element.

a. Normal operation and use consists of exterior envelope; pedestrian access & egress.
b. Limitations and restrictions to use for project application are n/a.

[56]Why Steel? Performance Study of Steel Doors and Frames Compared to Other Materials; October 2012; Sponsored by SDI, HMMA, CSDMA http://www.steeldoor.org/whysteel.php (accessed November 27, 2012).

c. Maintenance and repair are provided in accordance with the manufacturer's published recommendations and the BSLP maintenance, repair and replacement schedule. Special maintenance access requirements are fire-resistant doors to remain unobstructed.

Maintenance, Repair, and Replacement Schedule

a. Routine maintenance. Include maintenance instructions, recommended cleaning and accessory materials, and recommended maintenance frequency.

Door maintenance is associated primarily with the door accessories and hardware. The door should swing freely and smoothly from the opened to latched position. Inspect for loose screws, hinge pin wear and damage to sweeps, perimeter seals, and weatherstripping. Check glazing material for cracks and/or missing pieces of glazing. Inspect sweeps, perimeter seals and weatherstripping for splitting, cracking and deforming of flexible components. Gasketting should be positioned to create proper contact along the door's entire perimeter. The seal is inadequate if a sheet of paper, inserted between the door and gasket, is not firmly held in place when the door is closed. The paper test should be conducted along the entire perimeter. Install replacement components as needed.

For fire rated and/or smoke control assemblies provide and document regularly scheduled maintenance checks for hardware in accordance with SDI-124 Maintenance of Standard Steel Doors & Frames[57] and with local regulations. For additional guidance on inspection and maintenance of fire doors, refer to the Door and Hardware Institute (DHI) Owner's Guide: Swinging Fire Doors with Builder's Hardware—Annual Safety Inspections. The guide provides an overview of requirements of the NFPA 80,

[57]Steel Door Institute (SDI) publishes the SDI Technical Data Series. http://www.steeldoor.org/tech_data.php

Standard for Fire Doors and Other Opening Protectives. It explains the building owner's role in the care and maintenance of fire door assemblies.

Glazing: See B2020 Exterior Windows

Stainless Steel: See C1030 Fittings

b. Repair and replacement. Identify manufacturer support services for repair and replacement. These may include diagnostic services, parts availability, upgrade notification procedures, customer service contact information, and online technical support.

Perimeter sealants, properly designed and installed with high-quality material, have a reference service life of 10 to 20 years. For information on repair and replacement of steel doors and frames, refer to SDI-122 Installation and Troubleshooting Guide for Standard Steel Doors and Frames, and the SDI.

STEEL DOOR INSTITUTE (SDI)
30200 Detroit Road
Westlake, Ohio 44145
(440) 899-0010
www.steeldoor.org

For information on repair and replacement of door hardware, refer to the Door and Hardware Institute (DHI) bulletins and the DHI.

- ASD-1 Aluminum Storefront Doors
- BH-1 Butts and Hinges
- CH-1 Continuous Hinges
- EAH-91 Electrified Architectural Hardware
- ED-1 Exit Devices
- EL-1 Electromagnetic Locks
- FC-1 Concealed Floor Closers
- MK-1 Masterkeying

- P-1 Pivots
- PH-1 Protective Hardware
- SDC-92 Surface Door Closers
- WS-1/REV Gasketing and Thresholds

DOOR AND HARDWARE INSTITUTE

14150 Newbrook Drive, Suite 200

Chantilly, VA 20151-2232

(703) 222-2010

www.dhi.org

Adaptability, including limitations on and the potential for

a. Deconstruction and repurposing or reuse during the product design life.

   Doors are reusable. Doors are creatively repurposed into a variety of things, including, tables, headboards, wainscot, and anything that would require a door-sized board in construction.

b. Deconstruction and repurposing or recycling at end-of-life.

   Aluminum and Steel: See B2020 Exterior Windows

   Glazing: See B2020 Exterior Windows

   Stainless Steel: See C1030 Fittings

B3010 Roof Coverings

Name/Description

Roofing system

Classification

UNIFORMAT II—

B3010 Roof Coverings

MasterFormat—

07 41 11 Standing-Seam Metal Roof Panels

07 42 13.53 Metal Soffit Panels

07 54 19 Polyvinyl-Chloride (PVC) Roofing

07 62 00 Sheet Metal Flashing and Trim

Reference Service Life

Years: 20

Reference Service Life is equal to or greater than Building Design Life

Yes ____ No _✓_

Reference Service Life is equal to or greater than Building Element Design Life

Yes _✓_ No ____

Basis of determination:

____ Manufacturer's product data describing durability.

_✓_ Industry data or report(s) for durability compiled by government agency, academic institution, or a related professional or trade organization.

____ Reference service life established in accordance with ISO 15686 Part 2.

_✓_ Other information, including professional experience or observation of similar construction or materials in similar conditions, as identified by Architect/Engineer.

Describe:

Vinylroofs.org provides guidance on vinyl roofing. See below.[58] Polyvinyl chloride (PVC) roofing systems were developed in the 1960s. Many vinyl roof systems have demonstrated a service life in excess of 25 years. The expected lifespan is 20 to 30 years.

The warranty period specified for the roofing system on this project is 20 years from date of Substantial Completion. The warranty period is consistent with the minimum reference service life expectations from industry; therefore, the reference service life of PVC roofing is assessed at 20 years.

Coated steel elements typically are expected to have a long service life. See B1010 Floor Construction (Superstructure). However, exposure is greatest in the roof application. The warranty period specified for metal roofing is 20 years, including a 20 year warranty for the finish; therefore, the reference service life of the metal roofing is assessed at 20 years.

In-Use Conditions and Other Assumptions

Installation requirements. Installation of building element was in accordance with manufacturer's published instructions and project specifications.

Operation and use. Operation and use is normal for building element.

a. Normal operation and use consists of exterior envelope.
b. Limitations and restrictions to use for project application are

   Human traffic and activity (including installation and servicing of rooftop equipment) can produce cracks or punctures in the roofing membrane and lead to leaks. Care should be taken to avoid damage to roof from human traffic and activity. This may include advisory/warning signs and permanent or temporary walkways of rubber matting or other material as recommended by the roofing manufacturer.

---

[58]Vinylroofs.org is published by the Vinyl Roofing Division of the Chemical Fabrics and Film Association which was created to educate architects, specifiers, building owners and roofing contractors on the attributes of vinyl as a durable, reflective, heat-weldable material for single-ply roofing systems.

http://www.vinylroofs.org/index.html (accessed November 22, 2013).

PVC roof material has been known to shatter and crack in colder temperatures. Avoid human traffic and activity when the temperature outside is less than 50°F.

c. Maintenance and repair are provided in accordance with the manufacturer's published recommendations and the BSLP maintenance, repair and replacement schedule. Special maintenance access requirements are n/a.

Maintenance, Repair, and Replacement Schedule

a. Routine maintenance. Include maintenance instructions, recommended cleaning and accessory materials, and recommended maintenance frequency.

Inspect at least twice annually and after a storm event. Inspect for plugged drains, weathered sealants, and damaged flashing. Inspect vinyl roofing for pulled seams, cracks, punctures, or other indication of damage to the roofing membrane.

b. Repair and replacement. Identify manufacturer support services for repair and replacement. These may include diagnostic services, parts availability, upgrade notification procedures, customer service contact information, and online technical support.

PVC: According to Vinylroofs.org, damaged portions of the roof typically can be repaired without recoating or resurfacing. In most cases, the membrane can be cleaned and then repaired with the same materials and techniques used to weld the seams during installation.

PVC and coated metal: Where inspection indicates repair may be necessary, detailed evaluation and repair should be provided by a trained professional and in accordance with the system warranty.

Adaptability, including limitations on and the potential for

a. Deconstruction and repurposing or reuse during the product design life.

PVC: The industry does not support the reuse of PVC membrane at this time. For repurposing options, see end-of-life below.

Coated metal: See B1010 Floor Construction (Superstructure)

b. Deconstruction and repurposing or recycling at end-of-life.

PVC: Repurposing of PVC roofing is not prevalent; however, repurposing of PVC billboard membranes is gaining momentum. The repurposing options should be comparable and include pond liners, tarpaulins, boat covers, warehouse curtains, and tents.

PVC is the least recycled postconsumer plastic, per usage amount, in the United States. The U.S. Environmental Protection Agency (EPA) calculates that 910,000 tons of PVC waste are generated annually but less than a quarter of 1 percent is recovered for recycling.[59]

However, the feasibility and technical capability of recycling vinyl is established.[60] European roofing manufacturers have been recycling PVC roof membranes at the end of their service life for over a decade. Roofcollect®, the European collecting and recycling system, collected more than 1.3 million pounds of post-consumer vinyl in 2005. The collected material is processed and made available to manufacturers for incorporation into new vinyl roofing and waterproofing membranes. The Vinyloop® plant in Italy was inaugurated in 2002 and reportedly treats 10,000 tons of PVC scraps per year. It is also used for industrial research. In 2008, a new production line dedicated to the treatment of tarpaulins and other fiber containing PVC waste, was installed.

---

[59]http://www.epa.gov/climatechange/wycd/waste/downloads/Plastics.pdf (accessed November 22, 2013).

[60]A paper produced for Commission of the European Communities provides an overview of recycling options for PVC and outlines the potential benefits and challenges that may affect the continued development of recycling infrastructure for PVC roofing systems. http://eur-lex.europa.eu/LexUriServ/site/en/com/2000/com2000_0469en01.pdf (accessed November 22, 2013). See below.

Mechanical Recycling. In mechanical recycling, the PVC material is shredded and ground into a powder for use in new PVC products. The benefit of mechanically recycling PVC is that new products can achieve equal quality to originals (i.e., the material is not down-cycled).

Most PVC with recycled content has been mechanically recycled from industrial scrap material (http://eur-lex.europa.eu/LexUriServ/site/en/com/2000/com2000_0469en01.pdf, page 17).

Chemical recycling. In chemical recycling, PVC is broken down at a molecular level. At the molecular level, the chlorine and other chemical additives can be removed (and reclaimed for reuse) from the material. The challenge is that the process is specialized and costly.

As technology transfers from Europe to the United States, and as the market demand for PVC recycling increases, more options for reclamation and recycling should become available.

Resources for PVC roofing deconstruction, reuse, and recycling include

NATIONAL ROOFING CONTRACTORS ASSOCIATION
10255 W. Higgins Road; Suite 600
Rosemont, IL 60018-5607
(847) 299-9070
www.nrca.net

THE VINYL INSTITUTE
1737 King Street, Suite 390
Alexandria, VA, 22314
(571) 970-3400
www.vinylinfo.org

VINYLROOFS.ORG
www.vinylroofs.org
Produced by: Chemical Fabrics and Film Association
1300 Sumner Ave.
Cleveland, Ohio 44115-2851
(216) 241-7333
www.chemicalfabricsandfilm.com

Coated metal: See B1010 Floor Construction (Superstructure)

C1010 Interior Partitions

Name/Description

Interior partitions—metal stud/gypsum board partitions

Classification

UNIFORMAT II—

C1010 Interior Partitions

MasterFormat—

09 21 16.23 Gypsum Board Shaft Wall Assemblies

09 22 16 Non-Structural Metal Framing

09 29 00 Gypsum Board

Reference Service Life

Years: 50 for gypsum board, 100 for steel framing

Reference Service Life is equal to or greater than Building Design Life

Yes ✓ (steel) No ✓ (gypsum)

Reference Service Life is equal to or greater than Building Element Design Life

Yes ✓ No ____

Basis of determination:

____ Manufacturer's product data describing durability.

✓ Industry data or report(s) for durability compiled by government agency, academic institution, or a related professional or trade organization.

____ Reference service life established in accordance with ISO 15686 Part 2.

✓ Other information, including professional experience or observation of similar construction or materials in similar conditions, as identified by Architect/Engineer.

Describe:

The Gypsum Association states that under normal conditions, gypsum panels and gypsum plaster products have an expected service life in excess of 50 years.[61]

[61]Gypsum Association; http://www.gypsumsustainability.org/life-cycle.html (accessed November 12, 2012).

For coated steel elements, see B1010 Floor Construction (Superstructure)

In-Use Conditions and Other Assumptions

Installation requirements. Installation of building element was in accordance with manufacturer's published instructions and project specifications.

Operation and use. Operation and use is normal for building element.

a. Normal operation and use consists of interior, nonstructural.
b. Limitations and restrictions to use for project application are n/a.
c. Maintenance and repair are provided in accordance with the manufacturer's published recommendations and the BSLP maintenance, repair and replacement schedule. Special maintenance access requirements are n/a.

Maintenance, Repair, and Replacement Schedule

a. Routine maintenance. Include maintenance instructions, recommended cleaning and accessory materials, and recommended maintenance frequency.

   Gypsum: Gypsum board will be painted so requirements for routine cleaning will be primarily a function of the paint coating maintenance requirements. Enamel finishes may be lightly scrubbed with mild soap and water. Avoid saturating the paint coating and the gypsum board. Additionally, refer to the Gypsum Association (GA) Guidelines For Prevention of Mold Growth on Gypsum Board (GA-238-03).

   Steel framing will not be exposed and should not require routine maintenance.

b. Repair and replacement. Identify manufacturer support services for repair and replacement. These may include diagnostic services, parts availability, upgrade notification procedures, customer service contact information, and online technical support.

   Gypsum: Requirements for repair and replacement of finish materials applied to gypsum board should be provided in accordance with the recommendations of the finish material provider. For repair and replacement of gypsum board, refer to the recommendations of the Gypsum Association (GA), including the following[62]:

- Repair of Gypsum Board Joint Ridging (GA-221-2000). This publication describes the symptoms, causes, and repair methods for joint ridging in gypsum board surfaces.
- Repairing Screw or Nail Pops (GA-222-09). This publication describes the symptoms, causes, and repair methods for fastener pops in gypsum board surfaces.
- Repair of Fire-Rated Gypsum Board Systems (GA-225-08). This publication provides recommendations for repairing, patching, and replacing damaged fire-resistance-rated gypsum systems.
- Assessing Water Damage to Gypsum Board (GA-231-06). This publication provides information for assessing gypsum board that has been exposed to water or elevated levels of moisture and recommendations for drying.

Steel framing will not be exposed and should not require repair or replacement for the duration of the design life of the building.

Where renovation affects interior partitions, engage a trained professional to evaluate existing conditions and assess potential issues prior to renovation.

Adaptability, including limitations on and the potential for

a. Deconstruction and repurposing or reuse during the product design life. Gypsum: Generally, undamaged gypsum board may be removed and reused in similar applications. For reuse in fire-rated applications, verify acceptability for that purpose prior to use. The paper facing must remain

[62]Gypsum Association, www.gypsum.org (accessed November 14, 2012).

intact on gypsum board to be reused. Finish materials may be removed (so as not to damage the paper facing) or covered with new finish materials.

Scrap gypsum may be reclaimed onsite or recycled offsite. Refer to end-of-life below.

For coated steel elements, see B1010 Floor Construction (Superstructure)

b. Deconstruction and repurposing or recycling at end-of-life.

Gypsum: Typically, recycling of gypsum board has meant that "clean" construction scrap was crushed for lawns (i.e., soil amendment), industrial use and agricultural use. Markets for recycled gypsum include agriculture land applications (as a source of sulfur and calcium), compost additive, ingredient in portland cement, and source material for the production of new drywall. More recent reclamation technologies have targeted the mixed waste generated in renovation and demolition as well. The new technologies can handle gypsum board waste containing nails, screws, and various wall coverings.

Additional progress is anticipated for the gypsum recycling industry. EuroGypsum, the European association for the gypsum industry, has obtained a grant of 1.7 million Euro from the EU Life+ program for a project with the following title: "From cradle to cradle: a circular economy for the European gypsum industry with the demolition and recycling industry."[63] The research effort kicked off in January 2013 and will last for 3 years. The ultimate objective is to facilitate a closed-loop life cycle for gypsum. The more immediate aim of the project is to help the gypsum industry, and the related construction and waste sector, achieve the requirement of 70 percent recycling of gypsum waste in Europe by 2020 as prescribed by the EU Waste Directive.[64]

[63]www.gypsumtogypsum.org (accessed November 22, 2013).

[64]http://ec.europa.eu/environment/waste/framework/ (accessed November 22, 2013).

Resources for gypsum board deconstruction, reuse and recycling include

DRYWALLRECYCLING.ORG

Sponsored by the Construction & Demolition Recycling Association[65]

1585 Beverly Court, Ste. 112

Aurora, Illinois 60502

(630) 585-7530

www.drywallrecycling.org

GYPSUM RECYCLING AMERICA LTD

300 Centre Street

Holbrook, Massachusetts 02343

(617) 596-4297

www.gypsumrecycling.us

For coated steel elements, see B1010 Floor Construction (Superstructure)

[65]The Construction & Demolition Recycling Association (CDRA) was founded in 1996 as the Construction Materials Recycling Association (CMRA).

C1020 Interior Doors

Name/Description

Interior doors

Classification

UNIFORMAT II—

C1020 Interior Doors

MasterFormat—

08 11 13 Hollow Metal Doors and Frames

08 14 16 Flush Wood Doors

08 14 23 Aluminum-Clad Wood Doors

08 33 26 Overhead Coiling Grilles

08 71 00 Door Hardware

Reference Service Life

Years: 30

Reference Service Life is equal to or greater than Building Design Life

Yes ____ No _✓_

Reference Service Life is equal to or greater than Building Element Design Life

Yes _✓_ No ____

Basis of determination:

____ Manufacturer's product data describing durability.

_✓_ Industry data or report(s) for durability compiled by government agency, academic institution, or a related professional or trade organization.

____ Reference service life established in accordance with ISO 15686 Part 2.

_✓_ Other information, including professional experience or observation of similar construction or materials in similar conditions, as identified by Architect/Engineer.

Describe:

Research indicates an average of 30 years for interior doors such as wood, aluminum clad wood, and overhead coiling doors. National Institute of

Science and Technology (NIST) minimum reference service life of interior doors is 15 to 35 years, with painted coatings repaired/replaced every 3 to 10 years.[66] NIST averages for interior doors are 30+ years. NIST estimates for overhead garage doors are 20 to 50 years. Therefore, a reference service life for interior doors is assessed at 30 years.

In-Use Conditions and Other Assumptions

Installation requirements. Installation of building element was in accordance with manufacturer's published instructions and project specifications.

Operation and use. Operation and use is normal for building element.

a. Normal operation and use consists of access/egress for interior spaces.
b. Limitations and restrictions to use for project application are n/a.
c. Maintenance and repair are provided in accordance with the manufacturer's published recommendations and the BSLP maintenance, repair and replacement schedule. Special maintenance access requirements are n/a.

Maintenance, Repair, and Replacement Schedule

a. Routine maintenance. Include maintenance instructions, recommended cleaning and accessory materials, and recommended maintenance frequency.

   Door maintenance is associated primarily with the door accessories and hardware. The door should swing freely and smoothly from the opened to latched position. Inspect for loose screws and hinge pin wear. Check glazing material for cracks and/or missing pieces of glazing. Check finish for damage.

b. Repair and replacement. Identify manufacturer support services for repair and replacement. These may include diagnostic services, parts availability, upgrade notification procedures, customer service contact information, and online technical support.

   Hollow Metal Doors and Frames: See B2030 Exterior Doors

   Wood Doors: See C1030 Fittings.

[66]Refer to the appendix.

For information on the maintenance, repair, and replacement of motors, refer to Chapter 6.6, Motor Analysis, of Operations & Maintenance Best Practices: A Guide to Achieving Operational Efficiency, v2.0; Prepared by Pacific Northwest National Laboratory for the Federal Energy Management Program; U.S. Department of Energy.[67]

Resources for door maintenance, repair and replacement include

INTERNATIONAL DOOR ASSOCIATION
P.O. Box 246
West Milton, Ohio 45383-0246
(800) 355-4432
www.doors.org

WINDOW & DOOR MANUFACTURERS ASSOCIATION
2025 M Street, NW, Suite 800
Washington, D.C. 20036-3309
(202) 367-1157
www.wdma.com

Adaptability, including limitations on and the potential for

a. Deconstruction and repurposing or reuse during the product design life.

   Doors are reusable. Doors are creatively repurposed into a variety of things, including, tables, headboards, wainscot, and anything that would require a door-sized board in construction.

b. Deconstruction and repurposing or recycling at end-of-life.

   Aluminum and steel: The metal components of interior doors and frames (aluminum and steel) are considered to be recyclable. However,

[67]Operations & Maintenance Best Practices: A Guide to Achieving Operational Efficiency, v2.0; prepared by Pacific Northwest National Laboratory for the Federal Energy Management Program; U.S. Department of Energy; July 2004. http://wbdg.org/ccb/DOE/TECH/om_bestpractices_guide.pdf (accessed December 3, 2012).

The guide highlights O&M programs targeting 5 to 20 percent improvements in energy efficiency. It is intended to help achieve the intended service life of equipment in the facility.

the components are not typically separated during deconstruction but are removed together as mixed waste. Verify availability of recycling facilities for such mixed waste prior to deconstruction.

Wood: See C1030 Fittings

Controllers/operators may contain electronic waste. For deconstruction and repurposing of controllers/operators, refer to D30 HVAC.

C1030 Fittings

Name/Description

Furnishings and specialties (FFE)

Classification

UNIFORMAT II—

C1030 Fittings

MasterFormat—

06 10 00 Rough Carpentry

06 16 00 Sheathing

06 41 13 Wood-Veneer-Faced Architectural Cabinets

06 41 16 Plastic-Laminate-Faced Architectural Cabinets

06 46 00 Wood Trim

10 21 13 Toilet Compartments

10 28 00 Toilet, Bath and Laundry Accessories

12 36 40 Stone Countertops

Reference Service Life

Years: 100 years for stone; 60 years for wood; 40 years for stainless steel

Reference Service Life is equal to or greater than Building Design Life

Yes ✓ (stone and wood) No ✓ (stainless steel)

Reference Service Life is equal to or greater than Building Element Design Life

Yes ✓ No ___

Basis of determination:

___ Manufacturer's product data describing durability.

✓ Industry data or report(s) for durability compiled by government agency, academic institution, or a related professional or trade organization.

___ Reference service life established in accordance with ISO 15686 Part 2.

✓ Other information, including professional experience or observation of similar construction or materials in similar conditions, as identified by Architect/Engineer.

Describe:

Stone: See B2010 Exterior Walls

Wood: Wood is considered to be extremely durable. Empirical evidence and industry data indicate that a reference service life of 60+ years for wood construction is valid. Wood is resistant to high relative humidity and to many of the chemicals and conditions that adversely affect steel and concrete, such as corrosive salts, dilute acids, industrial stack gases and sea air. Pilings that are constantly submerged in fresh water have been known to last for centuries. The Empire State Building in New York and many of the structures in the city of Venice are built on wood foundation piles. A service life study that captured building age, building type, structural material, and reason for demolition for 227 buildings concluded that durability was comparable among wood frame, concrete and steel buildings. Sixty-three percent of the demolished wood buildings were older than 50 years at demolition and the majority were older than 75.[68]

A conservative determination assesses the reference service life for wood as being 60 years.

Stainless steel: The British Stainless Steel Association conducted a durability test for stainless steel in exterior applications. The test correlated durability with pitting. Results indicated that it took between 85 and 120 years to penetrate 1 mm in the stainless steel, depending on location (marine, semi-industrial, and rural) and by steel type (430, 304, and 316).[69]

The National Institute of Standards and Technology (NIST) average for the reference service life of stainless steel sinks is 40 years.[70]

A conservative determination assesses the reference service life for stainless steel as being 40 years.

---

[68]Survey on Actual Service Lives for North American Buildings—FPInnovations—Forintek Division; presented at Woodframe Housing Durability and Disaster Issues conference, Las Vegas, October 2004.
[69]http://www.bssa.org.uk/topics.php?article=51 (accessed December 4, 2012).
[70]Refer to the appendix.

In-Use Conditions and Other Assumptions

Installation requirements. Installation of building element was in accordance with manufacturer's published instructions and project specifications.

Operation and use. Operation and use is normal for building element.

a. Normal operation and use consists of office operations with limited public access/use.
b. Limitations and restrictions to use for project application are n/a.
c. Maintenance and repair are provided in accordance with the manufacturer's published recommendations and the BSLP maintenance, repair and replacement schedule. Special maintenance access requirements are n/a.

Maintenance, Repair, and Replacement Schedule

a. Routine maintenance. Include maintenance instructions, recommended cleaning and accessory materials, and recommended maintenance frequency.

   Stone: See B2010 Exterior Walls

   Wood: Maintenance of wood construction is largely related to protecting the wood from moisture. The Canadian Wood Council (CWC) provides guidance on the maintenance of wood construction. See information below.[71]

   Inspect annually for evidence of moisture damage. Inspect for discoloration of wood. Green stains may indicate algae growth. Brown stains may indicate failures in flashing. Inspect for decay above and below grade. Above grade, wood decays first below the wood surface (which dries faster). Evidence of above grade wood decay includes surface collapse, sunken paint, and cracking with softening. Below grade, wood decays first at the surface. Evidence of below grade decay includes softening and whitening or browning. Damp or musty smells may also be evidence of wood decay.

[71]www.durable-wood.com (accessed December 3, 2012). The durability site is a joint Canadian Wood Council (CWC)/FPInnovations project. The intent is to provide information that can help ensure the long service life of wood structures. CWC also publishes *Maintenance: Tips for Durable Wood Construction.*

Stainless steel: Stainless steel is a low carbon steel which contains chromium at 10 percent or more by weight. The chromium gives the steel its "stainless" properties, forming a corrosion-resisting chromium oxide film on the steel surface. Minor damage (mechanical or chemical) to the film should not be noticeable; the film is "self-healing" reacting with the oxygen in the air.

Stainless steel should require little routine maintenance. Provide maintenance in accordance with the Specialty Steel Industry of North America (SSINA) Handbook, *The Care and Cleaning of Stainless Steel.*[72] This guide is for both industrial and household stainless steel cleaning. It covers cleaning methods for routine cleaning, fingerprints and smears, stubborn stains and discoloration, grease, blood, burnt-on or baked-on foods, and grease and oil.

b. Repair and replacement. Identify manufacturer support services for repair and replacement. These may include diagnostic services, parts availability, upgrade notification procedures, customer service contact information, and online technical support.

Resources for wood maintenance, repair and replacement include

APA—THE ENGINEERED WOOD ASSOCIATION

7011 S. 19th Street

Tacoma, Washington 98466-5333

(253) 565-6600

www.apawood.org

BUILDING AND CONSTRUCTION TECHNOLOGY[73]

Room 125; Holdsworth Natural Resources Center

University of Massachusetts

---

[72]The Specialty Steel Industry of North America (SSINA) http://www.ssina.com/publications/index.html (accessed December 4, 2012).

[73]In 2009, the Building Materials and Wood Technology (BMATWT) center at the University of Massachusetts became the Building and Construction Technology center.

Amherst, Massachusetts 01003

(413) 545-1976

www.umass.edu/bmatwt/

CANADIAN WOOD COUNCIL

99 Bank Street, Suite 400

Ottawa, Ontario K1P 6B9

Canada

Helpdesk: (800) 463-5091 General: (613) 747-5544

www.cwc.ca

U.S. FOREST PRODUCTS LABORATORY

One Gifford Pinchot Drive

Madison, Wisconsin 53726

(608) 231-9200

www.fpl.fs.fed.us/

Adaptability, including limitations on and the potential for

a. Deconstruction and repurposing or reuse during the product design life.

Stone: See B2010 Exterior Walls

Wood: With care in deconstruction, wood elements can be reused. Wood building elements have been reused and repurposed for years. The reclaimed material may be cut and finished with same methods and standards as virgin material. Older lumber tends to be larger sections of tight-grained wood and is especially valued. Where wood elements are intended for reuse in structural applications, verify that the lumber meets applicable code requirements. It may be necessary to recertify and restamp the lumber.

The wood reclamation industry in the United States has evolved considerably since the 1970s and has an established network of service and distribution companies. Refer to the Directory of Wood-Framed

Building Deconstruction and Reused Building Materials Companies, 2005; U.S. Department of Agriculture—Forest Service.[74] This is a directory of companies involved in wood-framed building deconstruction, dismantling and reused building materials, with an emphasis on those that use, resell, and/or re-manufacture salvaged wood.

b. Deconstruction and repurposing or recycling at end-of-life.

Stone: See B2010 Exterior Walls

Wood: Wood products can be chipped and recycled for use as wood fiber. Markets for wood waste include feedstock for engineered woods, landscape mulch, soil conditioner, animal bedding, compost additive, sewage sludge bulking medium, and boiler fuel.

Wood is considered to be recyclable and recycling of wood fiber is common. However, wood building elements are not typically separated during deconstruction but rather are removed together with related components (e.g., fasteners, cladding, coatings) as mixed waste.[75] Verify availability of recycling facilities for such mixed waste prior to deconstruction.

Stainless steel: In most applications, stainless steel elements are easily separated and recycled. Stainless steel contains valuable raw materials such as iron, chromium and nickel which makes recycling stainless steel economically viable.

---

[74]Directory of Wood-Framed Building Deconstruction and Reused Building Materials Companies, 2005; U.S. Department of Agriculture—Forest Service; Forest Products Laboratory; General Technical Report FPL–GTR–150 http://www.doi.gov/greening/buildings/upload/The-Forest-Products-Laboratory-s-Directory-of-Wood-Framed-Building-Deconstruction-and-Reused-Building-Materials-Companies-fpl_gtr150.pdf.

[75]http://www.fpl.fs.fed.us/documnts/fplgtr/fplgtr133.pdf.

C2010 Stair Construction

Name/Description

Stairs

Classification

UNIFORMAT II—

C2010 Stair Construction

MasterFormat—

05 51 00 Metal Stairs

05 70 00 Decorative Metal

05 73 00 Decorative Metal Railings

Reference Service Life

Years: 100 years for coated steel elements.

Reference Service Life is equal to or greater than Building Design Life

Yes ✓ No ____

Reference Service Life is equal to or greater than Building Element Design Life

Yes ✓ No ____

Basis of determination:

____ Manufacturer's product data describing durability.

✓ Industry data or report(s) for durability compiled by government agency, academic institution, or a related professional or trade organization.

____ Reference service life established in accordance with ISO 15686 Part 2.

✓ Other information, including professional experience or observation of similar construction or materials in similar conditions, as identified by Architect/Engineer.

Describe: See B1010 Floor Construction (Superstructure)

In-Use Conditions and Other Assumptions

Installation requirements. Installation of building element was in accordance with manufacturer's published instructions and project specifications.

Operation and use. Operation and use is normal for building element.

a. Normal operation and use consists of vertical access/egress.

b. Limitations and restrictions to use for project application are n/a.

c. Maintenance and repair are provided in accordance with the manufacturer's published recommendations and the BSLP maintenance, repair and replacement schedule. Special maintenance access requirements are n/a.

Maintenance, Repair, and Replacement Schedule

a. Routine maintenance. Include maintenance instructions, recommended cleaning and accessory materials, and recommended maintenance frequency.

   See B1010 Floor Construction (Superstructure)

b. Repair and replacement. Identify manufacturer support services for repair and replacement. These may include diagnostic services, parts availability, upgrade notification procedures, customer service contact information, and online technical support.

   See B1010 Floor Construction (Superstructure)

Adaptability, including limitations on and the potential for

a. Deconstruction and repurposing or reuse during the product design life.

   See B1010 Floor Construction (Superstructure)

b. Deconstruction and repurposing or recycling at end-of-life.

   See B1010 Floor Construction (Superstructure)

C3010 Wall Finishes

Name/Description

Interior wall finishes

Classification

UNIFORMAT II—

C3010 Wall Finishes

MasterFormat—

09 24 00 Cement Plastering

09 30 00 Tiling

09 75 13 Stone Paneling

09 75 19 Stone Trim

Reference Service Life

Years: 66 for cement plaster; 100 for natural stone; 50 years for tile

Reference Service Life is equal to or greater than Building Design Life

Yes ✓ (plaster and stone) No ✓ (tile)

Reference Service Life is equal to or greater than Building Element Design Life

Yes ✓ No ____

Basis of determination:

____ Manufacturer's product data describing durability.

✓ Industry data or report(s) for durability compiled by government agency, academic institution, or a related professional or trade organization.

____ Reference service life established in accordance with ISO 15686 Part 2.

✓ Other information, including professional experience or observation of similar construction or materials in similar conditions, as identified by Architect/Engineer.

Describe:

Cement Plastering: See B2010 Exterior Walls

Stone: See B2010 Exterior Walls

Tile: The Tile Council of North America sponsored a Life Cycle Assessment of various floor finishes.[76] The study included reference service life statements for tile (quarry tile, glazed ceramic floor tile, glazed porcelain, mosaic, and unglazed porcelain) of 50 years. It included reference service life statements for portland cement terrazzo of 30 years and resin terrazzo of 15 years. An independent report from Dovetail[77] utilizing the Building for Energy and Environmental Sustainability (BEES) program of the National Institute of Standards and Technology (NIST)[78] assessed ceramic tile flooring at 50 years. As in-use conditions are more severe in flooring applications than in wall finishes, it is assumed that the flooring study represents a conservative value for the reference service life in wall finish applications. Therefore, the reference service life of tile is assessed as 50 years.

In-Use Conditions and Other Assumptions

Installation requirements. Installation of building element was in accordance with manufacturer's published instructions and project specifications.

Operation and use. Operation and use is normal for building element.

a. Normal operation and use consists of office operations with limited public access/use.
b. Limitations and restrictions to use for project application are n/a.
c. Maintenance and repair are provided in accordance with the manufacturer's published recommendations and the BSLP maintenance,

[76]In 2005, The Tile Council of North America commissioned an independent construction cost consulting firm, Scharf-Godfrey, to conduct a Life Cycle Cost study comparing various types of ceramic tile including glazed and unglazed, porcelain, mosaic, and quarry tile to twelve other floor finishes such as hardwood, laminate, concrete, stone, carpet, terrazzo, vinyl, and poured epoxy. http://www.tilethenaturalchoice.com/pdf/FloorCoveringComparison.pdf (accessed December 6, 2012).

[77]Dovetail Partners is a 501(c)(3) nonprofit organization that provides information about the impacts and tradeoffs of environmental decisions, including consumption choices, land use, and policy alternatives. http://www.dovetailinc.org/files/DovetailFloors0809.pdf (accessed December 6, 2012).

[78]The BEES program is accessible online (http://www.bfrl.nist.gov/oae/software/bees/) and free to download and use.

repair and replacement schedule. Special maintenance access requirements are n/a.

Maintenance, Repair, and Replacement Schedule

a. Routine maintenance: Include maintenance instructions, recommended cleaning and accessory materials, and recommended maintenance frequency.

Cement Plastering: See B2010 Exterior Walls

Stone: See B2010 Exterior Walls

Tile: Ceramic tile is nonporous and so is inhospitable to dust mites, mold, germs and bacteria. Additionally, several products today, including tiles and installation materials, have innovative coatings which inhibit microbial growth. Ceramic tile is inert. Routine maintenance includes wiping with gentle soaps. The Tile Council of North America provides guidelines for cleaning of tile surfaces and grout.

TILE COUNCIL OF NORTH AMERICA

100 Clemson Research Blvd.

Anderson, SC 29625

(864) 646-8453

www.tcnatile.com/

b. Repair and replacement. Identify manufacturer support services for repair and replacement. These may include diagnostic services, parts availability, upgrade notification procedures, customer service contact information, and online technical support.

Cement Plastering: See B2010 Exterior Walls

Stone: See B2010 Exterior Walls

Tile: The Tile Council of North America provides guidelines for repairing cracks and damaged tile installations.[79]

[79]See http://www.tcnatile.com/ and click on FAQs.

Adaptability, including limitations on and the potential for

a. Deconstruction and repurposing or reuse during the product design life.

Cement Plastering: See B2010 Exterior Walls

Stone: See B2010 Exterior Walls

Tile: With care in deconstruction, tile—especially adhesive set tile—can be reused. Removal of thin/thick set tile is likely to result in breakage so reuse may not be viable. Adhesive set tile often can be "popped" out without damage to the units. The reclaimed units may be cut and installed with the same methods as virgin material.

b. Deconstruction and repurposing or recycling at end-of-life.

Cement Plastering: See B2010 Exterior Walls

Stone: See B2010 Exterior Walls

Tile: Ceramic tile can be crushed and recycled for landscaping, as subbase material for pavements, or as aggregate for terrazzo or concrete. The industry is supportive of closed-loop life-cycle management and promotes Green Squared certification in accordance with ANSI A 138.1.[80] Green Squared certification evaluates several criteria including end-of-life management. Two credits are identified for end-of-life management.

- The first involves clean fill eligibility of a product. To satisfy this elective, a manufacturer provides documentation that the product is inert and solid such that it can qualify for state/local clean fill initiatives.
- The second involves end of product life collection plans. To satisfy this elective, the manufacturer establishes and implements a plan for the collection, processing, and recycling or repurposing of its products.

[80]ANSI A138.1 Green Squared—American National Standard Specifications for Sustainable Ceramic Tiles, Glass Tiles, and Tile Installation Materials.

Products eligible to achieve Green Squared certification include ceramic and glass tiles, grouts, mortars, backer boards, underlayments, and waterproof membranes.

As of 2012, recycling of postconsumer ceramic tile is not common. With the introduction of ANSI A138.1, it is anticipated that manufacturer take-back programs will expand as will the tile reclamation infrastructure.

Resources for tile deconstruction, reuse and recycling include

TILE COUNCIL OF NORTH AMERICA

100 Clemson Research Blvd.

Anderson, SC 29625

(864) 646-8453

www.tcnatile.com/

C3020 Floor Finishes

Name/Description

Floor finishes

Classification

UNIFORMAT II—

C3020 Floor Finishes

MasterFormat—

09 30 00 Tiling

09 64 00 Wood Flooring

09 66 23 Resinous Matrix Terrazzo Flooring

09 68 16 Sheet Carpeting

Reference Service Life

Years: 50 years for tile, wood, and terrazzo; 11 years for carpet

Reference Service Life is equal to or greater than Building Design Life

Yes ____ No ✓

Reference Service Life is equal to or greater than Building Element Design Life

Yes ✓ (tile, wood, terrazzo) No ✓ (carpet)

Basis of determination:

____ Manufacturer's product data describing durability.

✓ Industry data or report(s) for durability compiled by government agency, academic institution, or a related professional or trade organization.

____ Reference service life established in accordance with ISO 15686 Part 2.

✓ Other information, including professional experience or observation of similar construction or materials in similar conditions, as identified by Architect/Engineer.

Describe:

National Institute of Standards and Technology (NIST) minimums for service life values of floor surfaces are 3 to 20 years (undifferentiated).[81]

---

[81] Refer to the appendix.

NIST averages for service life values of floor surfaces are 20 to 30 years for vinyl, "lifetime" for wood, stone and terrazzo, and 11 years for carpeting. An independent report from Dovetail[82] utilizing the Building for Energy and Environmental Sustainability (BEES) program of the NIST assessed reference service life values for flooring materials as follows[83]:

- Carpet (nylon broadloom): 11 years
- Carpet (wool broadloom): 25 years
- Ceramic tile: 50 years
- Cork tile/plank: 50 years
- Linoleum: 30 years
- Terrazzo: 75 years
- Vinyl composition tile: 40 years

Proprietary information varies, but seems most consistent with the NIST and Dovetail studies and are in the 50-year range for all flooring except carpet. Therefore, the reference service life values are assessed as 50 years for tile, wood and terrazzo flooring; and, the reference service life of carpet is assessed as 11 years.

In-Use Conditions and Other Assumptions

Installation requirements. Installation of building element was in accordance with manufacturer's published instructions and project specifications.

Operation and use. Operation and use is normal for building element.

a. Normal operation and use consists of office operations with limited public access/use.
b. Limitations and restrictions to use for project application are n/a.
c. Maintenance and repair are provided in accordance with the manufacturer's published recommendations and the BSLP maintenance, repair and replacement schedule. Special maintenance access requirements are n/a.

[82]Dovetail Partners is a 501(c)(3) nonprofit organization that provides information about the impacts and tradeoffs of environmental decisions, including consumption choices, land use, and policy alternatives. http://www.dovetailinc.org/files/DovetailFloors0809.pdf (accessed December 6, 2012).
[83]The BEES program is accessible online (www.bfrl.nist.gov/oae/software/bees/) and free to download and use.

Maintenance, Repair, and Replacement Schedule

a. Routine maintenance. Include maintenance instructions, recommended cleaning and accessory materials, and recommended maintenance frequency.

Cleaning products should bear either a Green Seal[84] Certification Label and/or a U.S. EPA Design for the Environment (DfE)[85] label. Additionally, cleaning materials and methods should include the following:

- Mops and cloths should be microfiber to reduce the need for cleaning chemicals.
- Vacuum cleaners should meet the Carpet and Rug Institute (CRI) Seal of Approval Program—silver requirements.[86]
- Powered scrubbing machines should be equipped with a control method for variable rate dispensing to optimize the use of cleaning fluids.

Tile: See C3010 Wall Finishes

Wood: See C1030 Fittings

Terrazzo: The National Terrazzo and Mosaic Association provides guidance for terrazzo. See information below.[87]

Routine maintenance should include dry and damp mopping, with an occasional spray buffing. Harsh cleaners and sealers can damage terrazzo so use only materials that are neutral pH. Sealers should achieve a

---

[84]Green Seal; http://www.greenseal.org.

[85]EPA's Design for the Environment (DfE) label allows consumers to quickly identify and choose products that can help protect the environment and are safer for families. EPA's DfE program has allowed use of their logo on hundreds of products. In 2007, use of DfE labeled products reduced the use of "chemicals of concern" by 80 million pounds, up from 59 million pounds in 2006.

[86]The Carpet and Rug Institute (CRI) Green Label and Green Label Plus programs address carpeting and adhesives. CRI's Seal of Approval program addresses carpet cleaning products, methods and equipment. Under the Seal of Approval vacuum program, manufacturers must meet designated performance standards to earn gold, silver, or bronze ratings. http://www.carpet-rug.org/about-cri/cri-signature-programs.cfm (accessed November 26, 2013).

[87]The National Terrazzo and Mosaic Association; www.ntma.com.

slip resistance coefficient of friction rating as required by the applicable code. Water based sealers are most commonly recommended and utilized today.

Carpet: The Carpet and Rug Institute (CRI) provides guidance for carpet. See information below.[88]

Daily maintenance should include vacuuming and spot cleaning of all traffic areas.

- Vacuuming can remove more than 80 percent of dry soil on a daily basis. For heavily trafficked areas or where there is the potential for soil build-up, vacuuming should be done more frequently.
- Immediate cleaning of spots/spills will help reduce the probability of a permanent stain. Inspect carpet daily for spots/spills. Clean with CRI Seal of Approval certified carpet-cleaning solutions.

Monthly maintenance should include deep extraction cleaning to remove accumulated soil not removed by regular, daily vacuuming and spot removal.

b. Repair and replacement. Identify manufacturer support services for repair and replacement. These may include diagnostic services, parts availability, upgrade notification procedures, customer service contact information, and online technical support.

Tile: See C3010 Wall Finishes

Wood: See C1030 Fittings

Terrazzo: Where inspection indicates that repair or replacement of finished flooring may be necessary, engage a trained professional. The National Terrazzo and Mosaic Association provides a directory of manufacturers and contractors.

[88]http://www.carpet-rug.org/commercial-customers/cleaning-and-maintenance/index.cfm (accessed November 26, 2013).

Carpet: The Carpet and Rug Institute (CRI) provides guidance on the maintenance, repair, and replacement of carpet. Refer to CRI's *Troubleshooting Common Carpet Issues* publication.

Adaptability, including limitations on and the potential for

a. Deconstruction and repurposing or reuse during the product design life.

   Tile: See C3010 Wall Finishes

   Wood: See C1030 Fittings

   Terrazzo: Typically, repairs and refinishing can restore terrazzo flooring to its original luster. However, removal is likely to result in breakage so reuse may not be viable.

   Carpet: Reclaimed carpet may be cleaned and reinstalled in the same manner as virgin material. Reclaimed carpet is commonly repurposed for secondary flooring such as utilitarian padding, walk-off mats, and temporary floor protection.

b. Deconstruction and repurposing or recycling at end-of-life._

   Tile: See C3010 Wall Finishes

   Wood: See C1030 Fittings

   Terrazzo: Terrazzo is composed of naturally occurring aggregates, recycled glass or plastic and processed cement or epoxy binders. Terrazzo components can be crushed and recycled into aggregate for landscaping, as subbase material for pavements, or as aggregate for concrete. It can also be crushed into aggregate for use in new terrazzo flooring.

   Carpet: The reference service life of carpet is shorter than the targeted building element category; however, the carpet industry has developed a reclamation program that is supportive of closed-loop life-cycle management.

Carpet is a composite product made from face fibers that are bonded to primary and secondary backing material, usually with an adhesive. The most common face fibers (and their respective percentage of the carpet market) are: nylon (57 percent), polypropylene "Olefin" (36 percent), polyester "PET" (7 percent), and wool (0.4 percent). The primary challenge to recycling carpet has been separating the reclaimed carpet by face fiber type. Toward this goal, the Carpet and Rug Institute (CRI) introduced a Carpet Component Identification Code labeling system to make component identification easier for future recycling collection and sorting. This labeling system is currently being used by many CRI member companies. The labeling system is part of a broader industry effort, Carpet America Recovery Effort^{sm} (CARE), to develop market-based solutions for the recycling and reuse of postconsumer carpet.[89]

According to the Annual Report released April 2012 by CARE, members diverted 333 million pounds of carpet from the landfill in 2011. Of the carpet diverted, 250 million pounds were recycled back into carpet and other consumer products. From 2002 to 2011, CARE members have kept 2.3 billion pounds of waste carpet out of landfills.

[89]Carpet America Recover Effort (CARE) was established in January 2002 as a result of a Memorandum of Understanding for Carpet Stewardship (MOU), a national agreement signed by members of the carpet industry, representatives of government agencies at the federal, state and local levels, and nongovernmental organizations.

In 2011, CARE launched the CARE Certified Collectors Program. http://www.carpetrecovery.org/about.php (accessed November 26, 2013).

C3030 Ceiling Finishes

Name/Description

Ceiling finishes

Classification

UNIFORMAT II—

C3030 Ceiling Finishes

MasterFormat—

09 24 00 Cement Plastering

09 29 00 Gypsum Board

Reference Service Life

Years: 66 for cement plaster; 50 for gypsum board

Reference Service Life is equal to or greater than Building Design Life

Yes ✓ (plaster) No ✓ (gypsum)

Reference Service Life is equal to or greater than Building Element Design Life

Yes ✓ No _____

Basis of determination:

_____ Manufacturer's product data describing durability.

✓ Industry data or report(s) for durability compiled by government agency, academic institution, or a related professional or trade organization.

_____ Reference service life established in accordance with ISO 15686 Part 2.

✓ Other information, including professional experience or observation of similar construction or materials in similar conditions, as identified by Architect/Engineer.

Describe:

Cement Plastering: See B2010 Exterior Walls

Gypsum Board: See C1010 Interior Partitions

In-Use Conditions and Other Assumptions

Installation requirements. Installation of building element was in accordance with manufacturer's published instructions and project specifications.

Operation and use. Operation and use is normal for building element.

a. Normal operation and use consists of office operations with limited public access/use.
b. Limitations and restrictions to use for project application are n/a.
c. Maintenance and repair are provided in accordance with the manufacturer's published recommendations and the BSLP maintenance, repair and replacement schedule. Special maintenance access requirements are n/a.

Maintenance, Repair, and Replacement Schedule

a. Routine maintenance. Include maintenance instructions, recommended cleaning and accessory materials, and recommended maintenance frequency.

   Cement Plastering: See B2010 Exterior Walls

   Gypsum Board: See C1010 Interior Partitions

b. Repair and replacement. Identify manufacturer support services for repair and replacement. These may include diagnostic services, parts availability, upgrade notification procedures, customer service contact information, and online technical support.

   Cement Plastering: See B2010 Exterior Walls

   Gypsum Board: See C1010 Interior Partitions

Adaptability, including limitations on and the potential for

a. Deconstruction and repurposing or reuse during the product design life.

   Cement Plastering: See B2010 Exterior Walls

   Gypsum Board: See C1010 Interior Partitions

b. Deconstruction and repurposing or recycling at end-of-life.

   Cement Plastering: See B2010 Exterior Walls

   Gypsum Board: See C1010 Interior Partitions

D1010 Elevators and Lifts

Name/Description

Elevators

Classification

UNIFORMAT II—

D1010 Elevators And Lifts

MasterFormat—

14 20 50 General Elevator Requirements

14 21 00 Traction Elevators

14 24 00 Hydraulic Elevators

Reference Service Life

Years: 20

Reference Service Life is equal to or greater than Building Design Life

Yes ____ No ✓

Reference Service Life is equal to or greater than Building Element Design Life

Yes ✓ No ____

Basis of determination:

____ Manufacturer's product data describing durability.

✓ Industry data or report(s) for durability compiled by government agency, academic institution, or a related professional or trade organization.

____ Reference service life established in accordance with ISO 15686 Part 2.

✓ Other information, including professional experience or observation of similar construction or materials in similar conditions, as identified by Architect/Engineer.

Describe:

A report by APPA distinguishes the service life of elevators by life cycle of the driving machine, and life cycle of the controller as follows[90]:

[90]*Elevator Systems*; by Jay Popp; 2009; sponsored by APPA, www.appa.org; http://credentialing.appa.org/documents/BOKChapter21-ElevatorSystems.pdf (accessed December 6, 2012).

Elevator Type	Service Life of Driving Machine (years)	Service Life of Controller (years)
Hydraulic	20	20
Geared traction	30–35	20
Compact gearless traction (MRLs)	15–20	15
Gearless traction	50+	10–12

For the types of elevators used on this project, APPA reports a service life of the driving machine as 20 to 50+ years. A conservative determination assesses the reference service life for elevators on this project as being 20 years with an expectation that the controller may need repair or replacement in 10 to 20 years.

In-Use Conditions and Other Assumptions

Installation requirements. Installation of building element was in accordance with manufacturer's published instructions and project specifications.

Operation and use. Operation and use is normal for building element.

a. Normal operation and use consists of office operations with limited public access/use.
b. Limitations and restrictions to use for project application are n/a.
c. Maintenance and repair are provided in accordance with the manufacturer's published recommendations and the BSLP maintenance, repair and replacement schedule. Special maintenance access requirements are access to the operating components and controller for inspection and maintenance workers and equipment is required per ASME A17.1 Elevator Code.

Maintenance, Repair, and Replacement Schedule

a. Routine maintenance. Include maintenance instructions, recommended cleaning and accessory materials, and recommended maintenance frequency.

Inspections must be performed by qualified professionals at regular intervals as required by the authority having jurisdiction. Increasingly, jurisdictions are allowing authorized independent inspectors to perform periodic inspections that previously would have been performed by the government. In many states, a list of qualified inspection companies is available.[91] The ASME Qualified Elevator Inspector's (QEI) Accreditation Program covers organizations that certify elevator inspectors and inspector supervisors in accordance with the ASME QEI-1 Standard. Inspectors certified by accredited organizations are qualified to inspect elevators, escalators, and related equipment covered by the ASME A17.1 Elevator Code.[92]

ASME A17.1 Appendix N lists the recommended frequency for conveyance inspections and tests. Each authority having jurisdiction determines the frequency for inspection in its jurisdiction.

Resources for inspection and maintenance of elevators include

ELEVATOR INDUSTRY WORK PRESERVATION FUND
7154 Columbia Gateway Drive
Columbia, Maryland 21046
(410) 312-1474
EIWPFGWK@aol.com

NATIONAL ASSOCIATION OF ELEVATOR SAFETY
AUTHORITIES INTERNATIONAL
6957 Littlerock SW, Suite A
Tumwater, Washington 98512
(360) 292-4968
dotty@naesai.org

[91]The U.S. General Services Administration (GSA) maintains a list of independent elevator inspection service contractors. http://www.gsaelibrary.gsa.gov/ElibMain/sinDetails.do;jsessionid=CA5F291FBA2999EEB80ECB393D95BABB.node2?executeQuery=YES&scheduleNumber=03FAC&flag=&filter=&specialItemNumber=541+001 (accessed November 26, 2013).

[92]American Society of Mechanical Engineers (ASME) is discontinuing its QEI Accreditation Program. All accreditations will be withdrawn as of January 1, 2014. Refer to ASME QEI accredited organizations to locate certified inspectors. https://www.asme.org/shop/certification-accreditation/accreditation/elevator-inspector-certification-organization (accessed November 26, 2013).

QEI SERVICES, INC.

127 Union Avenue, Suite 2

Middlesex, New Jersey 08846

(888) 651-9209

info@qeiservices.com

b. Repair and replacement. Identify manufacturer support services for repair and replacement. These may include diagnostic services, parts availability, upgrade notification procedures, customer service contact information, and online technical support.

Where inspection indicates that repair or replacement is necessary, engage a qualified professional to provide evaluation and recommendations.

Adaptability, including limitations on and the potential for

a. Deconstruction and repurposing or reuse during the product design life.

For deconstruction and repurposing of elevator finishes, refer to the applicable finish section.

For deconstruction and repurposing of mechanical equipment and controllers, refer to D30 HVAC.

b. Deconstruction and repurposing or recycling at end-of-life.

For deconstruction and recycling of elevator finishes, refer to the applicable finish section.

For deconstruction and recycling of mechanical equipment and controllers, refer to D30 HVAC.

D20 Plumbing

Name/Description

Plumbing fixtures, equipment, and systems

Classification

UNIFORMAT II—

D20 Plumbing

MasterFormat—

22 05 00 Common Work Results for Plumbing

22 05 17 Sleeves and Sleeve Seals for Plumbing Piping

22 05 18 Escutcheons for Plumbing Piping

22 05 19 Meters and Gages for Plumbing Piping

22 05 23 General-Duty Valves for Plumbing Piping

22 05 29 Hangers and Supports for Plumbing Piping and Equipment

22 05 53 Identification for Plumbing Piping and Equipment

22 07 00 Plumbing Insulation

22 11 16 Domestic Water Piping

22 11 19 Domestic Water Piping Specialties

22 11 23 Domestic Water Pumps

22 13 16 Sanitary Waste and Vent Piping

22 13 23 Sanitary Waste Interceptors

22 14 23 Storm Drainage Piping Specialties

22 14 29 Sump Pumps

22 16 23 Natural Gas Piping

22 40 00 Plumbing Fixtures

22 46 00 Sub Surface Drainage

Reference Service Life

Years: 20

Reference Service Life is equal to or greater than Building Design Life

Yes ____ No _✓_

Reference Service Life is equal to or greater than Building Element Design Life

Yes _✓_ No ____

Basis of determination:

____ Manufacturer's product data describing durability.

_✓_ Industry data or report(s) for durability compiled by government agency, academic institution, or a related professional or trade organization.

____ Reference service life established in accordance with ISO 15686 Part 2.

_✓_ Other information, including professional experience or observation of similar construction or materials in similar conditions, as identified by Architect/Engineer.

Describe:

National Institute of Science and Technology (NIST) minimum reference service life of plumbing fixtures and equipment range from 5 to 25 years.[93] NIST minimum reference service life of piping range from 15 to 40 years. NIST averages for reference service life values of plumbing fixtures and equipment are 5 to 30 years depending on the component, and 50 to 100 years for waste piping. A study by the American Water Works Association Research Foundation assessed the reference service life of PVC pipe as 110+ years.[94] While major equipment is expected to last longer and minor items shorter, a reference service life for plumbing elements in general is assessed at 20 years. It is expected that minor items will be repaired or replaced, similar to sealants and gasketing in doors, in order to maintain the service life of the plumbing system as a whole.

[93]Refer to the appendix.

[94]http://www.uni-bell.org/environment.html (accessed December 10, 2012).

In-Use Conditions and Other Assumptions

Installation requirements. Installation of building element was in accordance with manufacturer's published instructions and project specifications.

Operation and use. Operation and use is normal for building element.

a. Normal operation and use consists of office operations with limited public access/use.
b. Limitations and restrictions to use for project application are n/a.
c. Maintenance and repair are provided in accordance with the manufacturer's published recommendations and the BSLP maintenance, repair and replacement schedule. Special maintenance access requirements are access to equipment, fixtures, valves, meters, clean-outs, etc.

Maintenance, Repair, and Replacement Schedule

a. Routine maintenance. Include maintenance instructions, recommended cleaning and accessory materials, and recommended maintenance frequency.

   To the greatest extent feasible, use cleaners that are gentle on the plumbing system and on the environment. Products should bear either a Green Seal[95] Certification Label and/or a U.S. EPA Design for the Environment (DfE)[96] label. Green Seal certified products include

   - Powdered Laundry Bleach (GC-11)
   - General-Purpose, Bathroom, Glass, and Carpet Cleaners Used For Household Purposes (GS-08)
   - Cleaning and Degreasing Agents (GS-34)
   - General-Purpose, Bathroom, Glass, and Carpet Cleaners Used for Industrial & Institutional Cleaners (GS-37)

[95] Green Seal; http://www.greenseal.org.

[96] EPA's Design for the Environment (DfE) label allows consumers to quickly identify and choose products that can help protect the environment and are safer for families. EPA's DfE program has allowed use of their logo on hundreds of products. In 2007, use of DfE labeled products reduced the use of "chemicals of concern" by 80 million pounds, up from 59 million pounds in 2006.

Routine inspection should include a visual review of piping, fixtures and equipment.

- Inspect exposed pipes for evidence of leaks such as drips, puddles, and watermarks or staining on adjacent surfaces.
- Inspect exposed pipes for evidence of corrosion such as green stains around brass and copper fittings and on shutoff valves, and yellow or orange stains on older steel pipe.
- Test water pressure: Low water pressure can be a sign of sediment build up in the faucet or shower head. Clean heads to remove sediment. If the pressure remains low, it may be a sign of a problem with the water line.
- Inspect drains for the speed of drainage: Sink drains should have a full swirl when draining. Slow drainage usually indicates a clog in the drain or a blocked vent pipe. Bubbles or gurgling sounds usually indicate a blocked vent pipe.
- Verify toilets flush properly: Verify that the water does not continue to run after flushing and that there is no water on the floor around the toilet.

Additionally, review complaints (if any) from building occupants. Various tastes and odors can be present in water, which can be traced to many different conditions. If objectionable tastes and/or odors are present, refer to the Water Quality Association and the local water utility for assistance in assessing the building's water quality.

WATER QUALITY ASSOCIATION
International Headquarters & Laboratory
4151 Naperville Road
Lisle, IL 60532-3696, USA
(630) 505 0160
www.wqa.org

b. Repair and replacement. Identify manufacturer support services for repair and replacement. These may include diagnostic services, parts availability, upgrade notification procedures, customer service contact information, and online technical support.

Where inspection indicates that repair or replacement is necessary, engage a qualified professional to provide evaluation and recommendations.

Resources for guidance on repair or replacement include

AMERICAN SOCIETY OF PLUMBING ENGINEERS

2980 S River Rd

Des Plaines, IL 60018

(847) 296-0002

www.aspe.org

PLUMBING-HEATING-COOLING CONTRACTORS

ASSOCIATION

180 S. Washington Street

Falls Church, VA 22046

(703) 237-8100; Toll Free: (800) 533-7694

www.phccweb.org

Adaptability, including limitations on and the potential for

a. Deconstruction and repurposing or reuse during the product design life.

Deconstruction for reuse is possible but may not be recommended. Where plumbing systems are deconstructed for system upgrades to adopt improved water management and/or energy management materials and methods, the removed systems represent older, less efficient technology. In such cases, it may be the better resource management option to recycle. Where plumbing systems are deconstructed and the fixtures are still considered to be efficient and functional items, reuse is preferred. The reclaimed components may be used in the same manner as virgin material.

b. Deconstruction and repurposing or recycling at end-of-life.

Fixtures: Ceramics can be crushed and recycled for landscaping, as sub-base material for pavements, or as aggregate for terrazzo or concrete.

Metal: Metals used in plumbing systems (e.g., iron, steel, and copper) are commonly recycled.

PVC: The PVC reclamation industry is growing globally.

In the European Union (EU), Vinyl 2010 and Vinylplus[97] set targets for recycling of PVC—800,000 tons/year of PVC by 2020, including 100,000 tons of difficult to recycle waste.[98] New technologies developed to meet the EU targets include the Vinyloop© process which uses an organic solvent to separate the PVC compound from other types of plastic waste or from the other materials in a PVC composite.[99]

In the United States, the Vinyl Institute provides information on recycling PVC, including the *Vinyl Recycling Directory*, which is a searchable database of companies that accept vinyl for processing.[100]

---

[97]http://www.vinylplus.eu/en_GB/about-vinylplus/our-voluntary-commitment (accessed December 10, 2012).

[98]In 2009, under Vinyl 2010, almost 83,000 tons of window profiles and other profiles were recycled; and almost 17,000 tons of PVC pipes were recycled (mostly back into pipes). As of 2012, collection programs exist in much of the European Union (EU). Schemes for pipes now cover nearly all "EU 15" countries, and Poland, Hungary, Czech Republic and Slovakia. http://www.pvc.org/en/p/pvc-recycling-by-application (accessed December 11, 2012).

[99]A special solvent is used to dissolve the PVC waste selectively in a closed loop batch process under pressure. The components of the original PVC formulation are recovered in the form of an aqueous suspension. http://www.pvc.org/en/p/pvc-recycling-by-application (accessed December 11, 2012).

[100]Vinyl Recycling Directory http://www.vinylinfo.org/recycling/directory/ (accessed November 26, 2013).

In addition, the Vinyl Institute maintains a directory of manufacturers that use recycled vinyl to make new products. http://www.vinylinfo.org/recycling/product-list/ (accessed November 26, 2013).

D30 Heating, Ventilating, and Air Conditioning (HVAC)

Name/Description

HVAC equipment and systems

Classification

UNIFORMAT II—

D30 HVAC

MasterFormat—

23 00 01 Mechanical Selections

23 05 00 Basic Mechanical Materials and Methods

23 05 13 Common Motor Requirements for HVAC Equipment

23 05 19 Meters and Gages

23 05 23 Valves

23 05 29 Hangers and Supports

23 05 53 Mechanical Identification

23 06 00 Refrigerant Monitoring Equipment

23 06 50 Variable Frequency Drives

23 07 00 Duct Insulation

23 07 10 HVAC Equipment Insulation

23 07 20 HVAC Piping Insulation

23 21 13 Hydronic Piping

23 21 15 Underground Hydronic Piping

23 21 23 Hydronic Pumps

23 25 02 HVAC Water Treatment

23 31 13 Metal Ducts

23 33 00 Duct Accessories

23 34 23 Power Ventilators

23 36 00 Air Terminal Units

23 64 16 Magnetic Bearing Chiller

23 65 00 Open Circuit Cooling Towers

23 65 02 Closed Circuit Cooling Towers

23 72 00 Air-to-air Energy Recovery Equipment

23 73 13 Central Air Handling Units

23 82 19 Fan Coil Units

Reference Service Life

Years: 20

Reference Service Life is equal to or greater than Building Design Life

Yes ____ No _✓_

Reference Service Life is equal to or greater than Building Element Design Life

Yes _✓_ No ____

Basis of determination:

____ Manufacturer's product data describing durability.

_✓_ Industry data or report(s) for durability compiled by government agency, academic institution, or a related professional or trade organization.

____ Reference service life established in accordance with ISO 15686 Part 2.

_✓_ Other information, including professional experience or observation of similar construction or materials in similar conditions, as identified by Architect/Engineer.

Describe:

Commissioning establishes project design requirements, including system design life and in-use conditions. Commissioning documents the normal operation and use as well as the maintenance and repair procedures for the design life of the building. See Commissioning Manual for the project.

National Institute of Standards and Technology (NIST) minimums for service life values of HVAC equipment range from 5 to 40 years depending on the component.[101] NIST average for service life values of HVAC equipment is 15 to 20 years. Therefore, a reference service life for HVAC elements in general is assessed at 20 years.

[101] Refer to the appendix.

In-Use Conditions and Other Assumptions

Installation requirements. Installation of building element was in accordance with manufacturer's published instructions and project specifications.

Operation and use. Operation and use is normal for building element.

a. Normal operation and use consists of

See Commissioning Manual for the project.

b. Limitations and restrictions to use for project application are

See Commissioning Manual for the project.

c. Maintenance and repair are provided in accordance with the manufacturer's published recommendations and the BSLP maintenance, repair and replacement schedule. Special maintenance access requirements are

See Commissioning Manual for the project.

Maintenance, Repair, and Replacement Schedule

a. Routine maintenance. Include maintenance instructions, recommended cleaning and accessory materials, and recommended maintenance frequency.

See Commissioning Manual for the project.

b. Repair and replacement. Identify manufacturer support services for repair and replacement. These may include diagnostic services, parts availability, upgrade notification procedures, customer service contact information, and online technical support.

See Commissioning Manual for the project.

Adaptability, including limitations on and the potential for

a. Deconstruction and repurposing or reuse during the product design life.

See Commissioning Manual for the project.

b. Deconstruction and repurposing or recycling at end-of-life.

Deconstruction for reuse is possible but may not be recommended. Where HVAC systems are deconstructed for system upgrades to adopt improved energy efficiency materials and methods, the removed systems represent older, less efficient technology. It such cases, it may be the better resource management option to recycle. Where HVAC systems are deconstructed and the equipment is still considered to be efficient and functional, reuse is preferred. The reclaimed components may be used in the same manner as virgin material. Electronics recyclers repair, refurbish and resell functioning electronics equipment as used products into domestic and international markets.

Metals used in HVAC systems are commonly recycled.

If components are identified for recycling, care should be taken to properly manage potentially hazardous materials. Hazardous materials are anticipated to include electronic waste and refrigerants.

Controllers/operators may contain electronic waste. Electronic waste, e-waste, e-scrap, or waste electrical and electronic equipment (WEEE) describes discarded electrical or electronic devices. The Consumer Electronics Association (CEA) promotes the recycling of end-of-life electronics through its recycling locator at www.GreenerGadgets.org[102] The emphasis of CEA is on consumer products such as mobile phones and computers; but the basic information can be useful in many cases. Refer to the U.S. Environmental Protection Agency (EPA) for information related to the management of e-waste and e-cycling.[103] Refer to the U.S. General Services Administration (GSA) for facilities management information regarding e-waste. In March 2012, the GSA announced new guidelines banning all federal agencies from disposing of electronic waste in landfills.[104] The policy directs U.S. federal agencies to reuse electronics to the maximum extent possible and to send

[102]The Consumer Electronics Association; www.GreenerGadgets.org (accessed December 3, 2012).

[103]http://www.epa.gov/wastes/conserve/materials/ecycling/index.htm (accessed November 25, 2013).

[104]http://www.gsa.gov/graphics/ogp/FMR_Bulletin_B-34.pdf (accessed December 3, 2012).

nonfunctioning products to certified e-waste recyclers, under R2 or eStewards, when reuse is not an option.[105]

Refrigerant handling and disposal is regulated. The regulations are likely to be modified (updated) during the service life of the building. Where handling or disposal of refrigerant is indicated, engage a qualified professional to provide evaluation and recommendations.

[105]In support of the federal initiative, the EPA encourages electronics recyclers to demonstrate that they meet specific safety and sustainability standards through accredited, third-party certification. Currently, two accredited certification standards exist: the Responsible Recycling Practices (R2) and the e-Stewards standards.

For information on the R2 standards and certification process, refer to R2 Solutions (R2S) which is a nonprofit organization established to house the R2 Practices. http://www.r2solutions.org/ (accessed November 25, 2013).

For information on the e-Stewards standards and certification process, refer to e-Stewards http://www.e-stewards.org (accessed November 25, 2013).

D40 Fire Protection

Name/Description

Fire protection system

Classification

UNIFORMAT II—

D40 Fire Protection

MasterFormat—

21 05 17 Sleeves and Sleeve Seals for Fire-Suppression Equipment

21 05 18 Escutcheons for Fire-Suppression Equipment

21 13 13 Wet-Pipe Sprinkler Systems

21 13 16 Dry-Pipe Sprinkler Systems

28 31 11 Digital, Addressable Fire-Alarm System

Reference Service Life

Years: 20

Reference Service Life is equal to or greater than Building Design Life

Yes ____ No ✓

Reference Service Life is equal to or greater than Building Element Design Life

Yes ✓ No ____

Basis of determination:

____ Manufacturer's product data describing durability.

✓ Industry data or report(s) for durability compiled by government agency, academic institution, or a related professional or trade organization.

____ Reference service life established in accordance with ISO 15686 Part 2.

✓ Other information, including professional experience or observation of similar construction or materials in similar conditions, as identified by Architect/Engineer.

Describe:

Sprinkler systems are considered to have a service life comparable to plumbing systems. Fire alarm systems are considered to have a service life comparable to electrical systems.

In-Use Conditions and Other Assumptions

Installation requirements. Installation of building element was in accordance with manufacturer's published instructions and project specifications.

Operation and use. Operation and use is normal for building element.

a. Normal operation and use consists of fire protection.
b. Limitations and restrictions to use for project application are n/a.
c. Maintenance and repair are provided in accordance with the manufacturer's published recommendations and the BSLP maintenance, repair and replacement schedule. Special maintenance access requirements are n/a.

Maintenance, Repair, and Replacement Schedule

a. Routine maintenance. Include maintenance instructions, recommended cleaning and accessory materials, and recommended maintenance frequency. See below.
b. Repair and replacement. Identify manufacturer support services for repair and replacement. These may include diagnostic services, parts availability, upgrade notification procedures, customer service contact information, and online technical support.

   Maintenance, repair and replacement should be provided in accordance with applicable regulatory requirements and NFPA standards and recommendations, including:

- NFPA 3, Recommended Practice on Commissioning and Integrated Testing of Fire Protection and Life Safety Systems
- Re-commissioning (Re-Cx) and Retro-commissioning (RCx) of Fire Protection and Life Safety Systems
- NFPA 25, Standard for the Inspection, Testing, and Maintenance of Water-Based Fire Protection Systems
- NFPA 297, Guide on Principles and Practices for Communications Systems
- NFPA 790, Standard for Competency of Third Party Field Evaluation Bodies

Resources for maintenance, repair and replacement include

BUILDING INSPECTION OFFICES

320 E. Jefferson Blvd.

Dallas, Texas 75203

www.dallascityhall.com/building_inspection/index.html

FIRESTOP CONTRACTORS INTERNATIONAL

ASSOCIATION

4415 W. Harrison St., #436

Hillside, IL 60162

(708) 202-1108

www.fcia.org

NATIONAL FIRE PROTECTION ASSOCIATION (NFPA)

1 Batterymarch Park

Quincy, Massachusetts, 02269

(617) 770-3000

www.nfpa.org

Adaptability, including limitations on and the potential for

a. Deconstruction and repurposing or reuse during the product design life.

   Plumbing: See D20 Plumbing

   Electrical: See D50 Electrical

b. Deconstruction and repurposing or recycling at end-of-life.

   Plumbing: See D20 Plumbing

   Electrical: See D50 Electrical

D50 Electrical

Name/Description

Electrical equipment and systems

Classification

UNIFORMAT II—

D50 Electrical

MasterFormat—

26 05 19 Low-Voltage Electrical Power Conductors and Cables

26 05 26 Grounding and Bonding for Electrical Systems

26 05 29 Hangers and Supports for Electrical Systems

26 05 33 Raceways and Boxes for Electrical Systems

26 05 44 Sleeves and Sleeve Seals for Electrical Raceways and Cabling

26 05 53 Identification for Electrical Systems

26 05 73 Overcurrent Protective Device Coordination Study

26 09 23 Lighting Control Devices

26 22 00 Low-Voltage Transformers

26 24 13 Switchboards

26 24 16 Panelboards

26 28 13 Fuses

26 28 16 Enclosed Switches and Circuit Breakers

26 29 13 Enclosed Controllers

26 32 13 Engine Generators

26 41 13 Lightning Protection for Structures

26 43 13 Transient-Voltage Suppression for Low-Voltage Electrical Power Circuits

26 51 00 Interior Lighting

26 56 00 Exterior Lighting

Reference Service Life

Years: 20

Reference Service Life is equal to or greater than Building Design Life

Yes ____ No _✓_

Reference Service Life is equal to or greater than Building Element Design Life

Yes _✓_ No ____

Basis of determination:

____ Manufacturer's product data describing durability.

_✓_ Industry data or report(s) for durability compiled by government agency, academic institution, or a related professional or trade organization.

____ Reference service life established in accordance with ISO 15686 Part 2.

_✓_ Other information, including professional experience or observation of similar construction or materials in similar conditions, as identified by Architect/Engineer.

Describe:

Commissioning establishes project design requirements, including system design life and in-use conditions. Commissioning documents the normal operation and use, and the maintenance and repair procedures for the design life of the building. See Commissioning Manual for the project.

National Institute of Science and Technology (NIST) minimum reference service life of electrical systems and equipment range from 5 to 40 years depending on the component.[106] NIST averages for service life values of electrical service and distribution is 30 years; lighting and branch wiring 20 years; special electrical systems such as diesel generator 25 years. While major equipment is expected to last longer and minor items shorter, a reference service life for electrical elements in general is assessed at 20 years. It is expected that minor items will be repaired or replaced, similar to sealants and gasketing in doors, in order to maintain the service life of the electrical system as a whole.

[106]Refer to the appendix.

In-Use Conditions and Other Assumptions

Installation requirements. Installation of building element was in accordance with manufacturer's published instructions and project specifications.

Operation and use. Operation and use is normal for building element.

a. Normal operation and use consists of

   See Commissioning Manual for the project.

b. Limitations and restrictions to use for project application are

   See Commissioning Manual for the project.

c. Maintenance and repair are provided in accordance with the manufacturer's published recommendations and the BSLP maintenance, repair and replacement schedule. Special maintenance access requirements are

   See Commissioning Manual for the project.

Maintenance, Repair, and Replacement Schedule

a. Routine maintenance. Include maintenance instructions, recommended cleaning and accessory materials, and recommended maintenance frequency.

   See Commissioning Manual for the project.

b. Repair and replacement. Identify manufacturer support services for repair and replacement. These may include diagnostic services, parts availability, upgrade notification procedures, customer service contact information, and online technical support.

   See Commissioning Manual for the project.

Adaptability, including limitations on and the potential for

a. Deconstruction and repurposing or reuse during the product design life.

   See Commissioning Manual for the project.

b. Deconstruction and repurposing or recycling at end-of-life.

   Metals used in electrical distribution systems are commonly recycled.

   Electrical equipment may be reclaimed for refurbishment and reuse. Electronics recyclers repair, refurbish and resell functioning electronic equipment as used products into domestic and international markets. In 2011, the U.S. electronics recycling industry processed over 4 million tons of used electronics equipment, producing specification grade commodities (e.g., steel, aluminum, copper, lead, plastics and glass).[107]

If components are identified for recycling, care should be taken to properly manage potentially hazardous materials. Engage a qualified professional.

Controllers/operators may contain electronic waste. For deconstruction and repurposing of controllers/operators, refer to D30 HVAC.

[107]Institute of Scrap Recycling Industries, Inc., Electronics Fact Sheet; 4/2013.

G2030 Pedestrian Paving

Name/Description

Pedestrian pavement and site hardscape

Classification

UNIFORMAT II—

G2030 Pedestrian Paving

MasterFormat—

32 13 13 Pedestrian Sidewalks

32 14 00 Unit Pavers

Reference Service Life

Years: 40

Reference Service Life is equal to or greater than Building Design Life

Yes ____ No ✓

Reference Service Life is equal to or greater than Building Element Design Life

Yes ✓ No ____

Basis of determination:

____ Manufacturer's product data describing durability.

✓ Industry data or report(s) for durability compiled by government agency, academic institution, or a related professional or trade organization.

____ Reference service life established in accordance with ISO 15686 Part 2.

✓ Other information, including professional experience or observation of similar construction or materials in similar conditions, as identified by Architect/Engineer.

Describe:

The National Concrete Pavement Technology Center states that service life for concrete pavement is 40+ years for low maintenance pavement. Industry continues to explore technologies that extend service lives.[108] Most service life analysis of concrete pavement is conducted on roadway pavements. As

[108]International Conference on Long-Life Concrete Pavements—2012; September 18–21, 2012—Seattle, Washington, http://www.cptechcenter.org/t2/documents/FHWAACPTConfon-LongLifeConcretePavements-Seattle-Sept2012-v081611web.pdf (accessed December 12, 2012).

in-use conditions are more severe in vehicular pavements than in pedestrian pavements, it is assumed that the vehicular pavement value represents a conservative reference service life value for pedestrian pavement.

The brick industry uses 40 years as their basis of determining life-cycle costs of clay pavers.[109] This is considered a conservative value. Clay brick pavers have been used in the United States since colonial times, with many clay paver systems still functional after 100+ years.

Therefore, the service life value of pedestrian paving (both concrete paving and clay brick pavers) is assessed as 40 years.

In-Use Conditions and Other Assumptions

Installation requirements. Installation of building element was in accordance with manufacturer's published instructions and project specifications.

Operation and use. Operation and use is normal for building element.

a. Normal operation and use consists of exterior pedestrian traffic.
b. Limitations and restrictions to use for project application are n/a.
c. Maintenance and repair are provided in accordance with the manufacturer's published recommendations and the BSLP maintenance, repair and replacement schedule. Special maintenance access requirements are n/a.

Maintenance, Repair, and Replacement Schedule

a. Routine maintenance. Include maintenance instructions, recommended cleaning and accessory materials, and recommended maintenance frequency.

   Concrete pavement: The National Concrete Pavement Technology Center provides guidance on concrete paving. See information below.[110]

[109]Brick Industry Association, www.gobrick.com (accessed December 12, 2012).

[110]The National Concrete Pavement Technology Center, http://www.cptechcenter.org/.

The Center offers numerous technical notes and publications, including *Sustainable Concrete Pavements: A Manual of Practice*; January 2012; Institute of Transportation, Iowa State University Research Park; Ames, IA; technical editors: Dr. Peter Taylor, Dr. Tom Van Dam; http://www.cptechcenter.org/technical-library/documents/Sustainable_Concrete_Pavement_508.pdf (accessed December 4, 2013).

When pavement shows evidence of deterioration (such as uneven settling, cracks, chips, and spalling), preventative maintenance techniques may be used. These include

- Slab stabilization
- Partial-depth repair
- Full-depth repair
- Use of precast panels in full-depth repairs
- Retrofitted edge drains
- Load transfer restoration
- Diamond grinding and grooving
- Joint resealing and crack sealing

Preventative maintenance techniques do not improve the structural capacity of the pavement. When the structural capacity of the pavement needs to be improved, rehabilitation techniques, including the use of concrete overlays, are necessary. See repair and replacement below.

Where inspection indicates that preventative and/or rehabilitation techniques are necessary, engage a qualified professional to provide evaluation and recommendations.

Unit pavers: The Brick Industry Association provides guidance on clay brick. See information below.[111]

Clay pavers are resistant to absorption of stains and can be kept clean in most environments by regular sweeping. Cleaning of brick pavements is similar to cleaning vertical brickwork. See B2010 Exterior Walls.

Mortar-filled joints generally are more resistant to aggressive cleaning methods (i.e., pressure washers). Sand-filled joints subjected to aggressive cleaning methods should contain stabilized joint sand or should be treated with a joint sand stabilizer.

[111]Brick Industry Association, www.gobrick.com.

Efflorescence may form on the paver surface due to soluble compounds in the underlying soils or to deicing chemicals, which remain as absorbed water evaporates from the pavement surface. Efflorescence may wear away with traffic and weathering, or may be brushed off dry pavers. Persistent efflorescence may indicate problems with subsurface drainage.

For deicing, BIA recommends calcium magnesium acetate and urea. Rock salt contains calcium chloride and can cause efflorescence. Sand rubbing on pavers with traffic can grind the paver surface. Sand used to provide traction on ice should be removed as soon as possible when the weather improves.

b. Repair and replacement. Identify manufacturer support services for repair and replacement. These may include diagnostic services, parts availability, upgrade notification procedures, customer service contact information, and online technical support.

Concrete pavement: Pavement rehabilitation usually involves the overlay of additional pavement in varying thicknesses depending on the degree to which structural capacity needs to be increased. When pavement deteriorates to the point of failure, it should be deconstructed and replaced.

Unit pavers: Pavers that are broken or chipped more than 10 percent of their cumulative edges should be replaced to maintain a level surface.

Resources for maintenance repair and replacement include

AMERICAN CONCRETE PAVEMENT ASSOCIATION

5420 Old Orchard Road

Skokie, Illinois 60077

(847) 966-2272

www.pavement.com

NATIONAL CONCRETE PAVEMENT TECHNOLOGY CENTER

CP Tech Center

Iowa State University

2711 South Loop Drive, Suite 4700

Ames, Iowa 50010-8664

(515) 294-5798

www.cptechcenter.org

THE BRICK INDUSTRY ASSOCIATION

1850 Centennial Park Drive, Suite 301

Reston, Virginia 20191

(703) 620-0010

www.gobrick.com

THE NATIONAL BRICK RESEARCH CENTER

Clemson University

100 Clemson Research Boulevard

Anderson, South Carolina 29625

(864) 656-1094

www.brickandtile.org/

Adaptability, including limitations on and the potential for

a. Deconstruction and repurposing or reuse during the product design life.

   Concrete pavement: See A1010 Standard Foundations

   Sections of concrete pavement may be repurposed for use in landscape and riprap.

   Unit pavers: See B2010 Exterior Walls

b. Deconstruction and repurposing or recycling at end-of-life.

   Concrete pavement: See A1010 Standard Foundations

   In-situ recycling of concrete pavement into new concrete pavement (i.e., crushed for use as subbase aggregate or in the bottom lift of a two-lift paving system[112]), offers the economic and environmental advantages of minimal material transportation.

   Unit pavers: See B2010 Exterior Walls

[112]Two-lift pavement system consist of a thick bottom lift (usually 80 percent or more of the total thickness) and a thin (20 percent or less of the total thickness) top lift that is optimized for the anticipated traffic. Two-lift concrete pavement is an established process. The oldest concrete pavements in the United States, constructed by the R.S. Blome Granitoid Co., were two-lift systems.

G2040 Site Improvements

Name/Description

Site improvements

Classification

UNIFORMAT II—

G2040 Site Development—Fences, Retaining Walls, Fountains And Other Site Improvements

MasterFormat—

32 84 00 Planting Irrigation

32 92 00 Landscape Planting

Reference Service Life

Years: 80 for landscaping; 20 for irrigation

Reference Service Life is equal to or greater than Building Design Life

Yes ✓(landscaping) No ✓(irrigation)

Reference Service Life is equal to or greater than Building Element Design Life

Yes ✓(landscaping) No ✓(irrigation)

Basis of determination:

_____ Manufacturer's product data describing durability.

__✓__ Industry data or report(s) for durability compiled by government agency, academic institution, or a related professional or trade organization.

_____ Reference service life established in accordance with ISO 15686 Part 2.

__✓__ Other information, including professional experience or observation of similar construction or materials in similar conditions, as identified by Architect/Engineer.

Describe:

Irrigation: The American Society for Golf Course Architects states that irrigation service life is 10 to 30 years.[113] The U.S. EPA GreenScapes

[113]http://www.asgca.org/articles/207-making-your-case-for-an-irrigation-renovation (accessed December 4, 2013).

program utilizes a service life of 20 to 25 years in its subsurface drip irrigation cost calculator.[114] Professional experience indicates that irrigation pipe and fittings have a service life of 20+ years, while the heads and valves have a service life of 5 to 25 years. Allowing for replacement of exposed components, the irrigation system service life is assessed as 20 years.

Landscaping: Landscaping is designed with xeriscaping principles. Plants are selected to be appropriate to the climate such that, after establishment, they will thrive naturally, with little maintenance or water. The natural lifespan of the tree and shrub species on this project range from 80 to 250 years. Grasses and flowers are perennials. The reference service life of landscaping is conservatively assessed as 80 years.

In-Use Conditions and Other Assumptions

Installation requirements. Installation of building element was in accordance with manufacturer's published instructions and project specifications.

Operation and use. Operation and use is normal for building element.

a. Normal operation and use consists of exterior site; pedestrian access.
b. Limitations and restrictions to use for project application are n/a.
c. Maintenance and repair are provided in accordance with the manufacturer's published recommendations and the BSLP maintenance, repair and replacement schedule. Special maintenance access requirements are n/a.

Maintenance, Repair, and Replacement Schedule

a. Routine maintenance. Include maintenance instructions, recommended cleaning and accessory materials, and recommended maintenance frequency.

   Irrigation: An inspection of the irrigation system is recommended after one year of operation and at least every three years thereafter. The inspection should be conducted by a qualified auditor, such as a U.S. EPA WaterSense partner.[115]

---

[114]http://www.epa.gov/epawaste/conserve/tools/greenscapes/tools/drip.pdf (accessed December 4, 2013).

[115]For a listing of EPA WaterSense irrigation partners by state, see www.epa.gov/watersense/pp/lists/irr_partners.htm.

The Whole Building Design Guide provides guidance on irrigation inspection. See information below.[116]

Inspection of the irrigation system should include

- General performance: Assess system performance; verify proper scheduling; identify deficiencies including deficiencies due to damage or modification of system, growth of landscape, or an aging system; identify opportunities to employ new technologies.
- Maintenance: Review maintenance documentation.
- Leaks: Check for leaks in system.
- Runoff/overspray: Check for runoff and overspray.
- Distribution uniformity: Irrigation systems should achieve a lower quarter distribution uniformity (DULQ) of 70 percent or greater. Measure distribution uniformity.

Irrigation should be winterized where there is danger of freezing. Winterizing should include the following steps:

- Flush the backflow assembly to protect internal components and brass body.
- Purge the irrigation system of water to protect external components from freezing and cracking.
- Shut off the main valve to prevent water from seeping into the system.

Landscaping: Xeriscaping utilizes indigenous plants, and low maintenance plants which are tolerant of the site's existing soils and climate without supplemental irrigation or fertilization once established.

[116]The Whole Building Design Guide (WBDG), a program of the U.S. National Institute of Building Sciences, includes the Federal Green Construction Guide for Specifiers. Section 32 84 00 provides recommendations for regular inspections. http://www.wbdg.org/design/greenspec.php (accessed December 4, 2013).

Every season, landscaping should be inspected visually to identify problems in plant health and growth. Where problems are identified, a soil analysis may be useful. The analysis should be performed by a laboratory with experience in such testing and recognized by the State Department of Agriculture.

The soil analysis should assess soil conditions to determine if they are appropriate for the plants; and, provide recommendations for nutrients and amendments as necessary. It should report percentages of organic matter; gradation of sand, silt, and clay content; cation exchange capacity; sodium absorption ratio; deleterious material; pH; and mineral and plant-nutrient content of topsoil.

b. Repair and replacement. Identify manufacturer support services for repair and replacement. These may include diagnostic services, parts availability, upgrade notification procedures, customer service contact information, and online technical support.

Irrigation: Typically, the average sprinkler system requires repairs or replacements within 5 to 10 years. This includes sprinkler heads, sprinkler valves, controllers, rain sensors, and pipe repairs.

If plants grow and block the spray from the sprinkler heads, sprinklers heads either need to be added to improve coverage, raised above plantings, or moved to provide better coverage.

Resources for maintenance, repair and replacement include

IRRIGATION ASSOCIATION

6540 Arlington Blvd.

Falls Church, VA 22042

(703) 536-708.

www.irrigation.org

The Irrigation Association publishes "Turf and Landscape Irrigation Best Management Practices" for guidance on maintenance, repair and replacement.[117]

Landscaping: Refer to landscaper's literature in Owner's O&M manual. Resources for maintenance, repair and replacement include

AMERICAN NURSERY AND LANDSCAPE ASSOCIATION
1200 G Street NW, Suite 800
Washington, DC 20005
(202) 789-2900
www.anla.org

ASSOCIATION OF PROFESSIONAL LANDSCAPE DESIGNERS
4305 North Sixth Street, Suite A
Harrisburg, PA 17110
(717) 238-978
http://apld.org

ECOLOGICAL LANDSCAPING ASSOCIATION
841 Worcester Road #326
Natick, MA 01760
(617) 436-5838
www.ecolandscaping.org

AMERICAN SOCIETY OF LANDSCAPE ARCHITECTS
636 Eye Street, NW
Washington, DC 20001-3736
(202) 898-2444; Tool Free: (888) 999-ASLA (2752)
www.asla.org

[117]www.irrigation.org/uploadedFiles/Resources/BMP_Revised_12-2010.pdf (accessed December 13, 2012).

In November 2009, the Sustainable Sites Initiative released a rating system for the design, construction and maintenance of sustainable landscapes, with or without buildings.[118]

U.S. ENVIRONMENTAL PROTECTION AGENCY (EPA)

www.epa.gov

The EPA GreenScapes program provides cost-efficient and environmentally friendly solutions for landscaping design, construction and maintenance.[119]

Adaptability, including limitations on and the potential for

a. Deconstruction and repurposing or reuse during the product design life.

   Irrigation: See D20 Plumbing

   Landscaping: Plants may be removed and replanted in another location. Even large plants and trees may be relocated with proper care. Relocation is stressful on a plant. Where relocation is desired, engage a professional landscaper or arborist.

b. Deconstruction and repurposing or recycling at end-of-life.

   Irrigation: See D20 Plumbing

   Landscaping: The Whole Building Design Guide provides guidance on composting. See information below.[120]

   Recycling of plants may be accomplished on site or off with mulching or composting. For chipping to use as mulch, identify organic debris that is free of disease, pest infestation, and chemical contamination. For

[118]http://www.sustainablesites.org/report/ (accessed December 13, 2012).

[119]http://www.epa.gov/epawaste/conserve/tools/greenscapes/index.htm (accessed February 27, 2014).

[120]The Whole Building Design Guide (WBDG), a program of the U.S. National Institute of Building Sciences, includes the Federal Green Construction Guide for Specifiers. Section 01 74 19 provides recommendations for composting site debris. Section 31 25 73 provides recommendations for use of compost in storm water management. http://www.wbdg.org/design/greenspec.php (accessed December 4, 2013).

composting on site, create composting containers or areas in accordance with State Extension Service recommendations and as follows:

- Moisture content: Maintain between 35 and 60 percent.
- Carbon to nitrogen (C/N) ratio: Maintain at approximately 30 to 1 by weight.
- Do not compost meat or dairy products on site.
- Avoid composting diseased vegetation on site. Because the operations of commercial composting facilities are monitored and controlled to maintain the high temperatures required in the thermophyllic phase of composting (the thermophyllic phase of the average residential composting pile is only 4 to 7 days and relies on thermophyllic bacteria that function at 104° to 170°F and that are extremely efficient at processing compostables) commercial composting can accept diseased vegetation and, in many cases, all types of animal waste. However, composting on site is not generally so well controlled. If temperatures in the pile do not get hot enough to kill the undesirable organisms, those organisms can reinfest new vegetation when the compost is applied.

Master Composter training programs are offered in many locations. Master Composter programs are designed to train and recruit local volunteers and to promote backyard composting.

Resources for information on composting include

TEXAS COMMISSION ON ENVIRONMENTAL QUALITY

12100 Park 35 Circle

Austin, TX 78753

www.tceq.texas.gov/

G3030 Storm Sewer

Name/Description

Storm water management

Classification

UNIFORMAT II—

G3030 Storm Sewer

MasterFormat—

02500 Storm Drainage System

02668 Water Improvements

02730 Sanitary Sewerage

Reference Service Life

Years: 75

Reference Service Life is equal to or greater than Building Design Life

Yes ✓ No ____

Reference Service Life is equal to or greater than Building Element Design Life

Yes ✓ No ____

Basis of determination:

____ Manufacturer's product data describing durability.

✓ Industry data or report(s) for durability compiled by government agency, academic institution, or a related professional or trade organization.

____ Reference service life established in accordance with ISO 15686 Part 2.

✓ Other information, including professional experience or observation of similar construction or materials in similar conditions, as identified by Architect/Engineer.

Describe:

National Institute of Science and Technology (NIST) minimum reference service life values of waste piping are 50 to 100 years.[121] A study by the

[121] Refer to the appendix.

American Water Works Association Research Foundation put the life expectancy of PVC pipe at more than 110 years.[122] The reference service life of concrete is conservatively assessed as 66 years (see A1010 Standard Foundations). The reference service life for storm sewerage is assessed as 75 years, representing the midpoint of the NIST averages and a value reasonably consistent with material specific data.

In-Use Conditions and Other Assumptions

Installation requirements. Installation of building element was in accordance with manufacturer's published instructions and project specifications.

Operation and use. Operation and use is normal for building element.

a. Normal operation and use consists of office operations with limited public access/use.
b. Limitations and restrictions to use for project application are n/a.
c. Maintenance and repair are provided in accordance with the manufacturer's published recommendations and the BSLP maintenance, repair and replacement schedule. Special maintenance access requirements are cleanouts/manholes.

Maintenance, Repair, and Replacement Schedule

a. Routine maintenance. Include maintenance instructions, recommended cleaning and accessory materials, and recommended maintenance frequency.

   See D20 Plumbing

b. Repair and replacement. Identify manufacturer support services for repair and replacement. These may include diagnostic services, parts availability, upgrade notification procedures, customer service contact information, and online technical support.

   See D20 Plumbing

Adaptability, including limitations on and the potential for

a. Deconstruction and repurposing or reuse during the product design life.

   Concrete: See A1010 Standard Foundations

[122] http://www.uni-bell.org/environment.html (accessed December 10, 2012).

Metal: See D20 Plumbing

PVC: See D20 Plumbing

b. Deconstruction and repurposing or recycling at end-of-life.

Concrete: See A1010 Standard Foundations

Metal: See D20 Plumbing

PVC: See D20 Plumbing

PART 4

Appendix

The following text and tables are excerpted from a National Institute of Standards and Technology (NIST) building materials study.[123] The study was sponsored through the PATH program and so focused on housing durability.[124] However, it should be noted that the study looked at building elements rather than the whole building or building system. Because of the focus on building elements, much of the information is applicable to commercial construction as well.

Baseline Measures of Service Life for Key Building Elements

The three tables presented in this section provide a detailed snapshot of service life data for typical building elements. Because service life data reported in the three tables come from different sources, there are differences both in data values and the elements reported. All three tables use the modified UNIFORMAT II elemental classification. Differences in data values stem from differences in focus (i.e., minimum service life versus average service life) and maintenance procedures (i.e., preventive versus reactive). Differences in building elements for which service life data are reported occur primarily at level 3—the most disaggregated and most detailed level of the elemental classification.

Table 4-1 presents service life data published in ASTM Standard Guide E 2136. These service life data are organized within a hierarchy that has ten level 1 elements. The ten level 1 elements are (1) substructure,

[123]Baseline Measures for Improving Housing Durability; NISTIR 6870; Robert E. Chapman and Christine A. Izzo; Sponsored by: National Institute of Standards and Technology, Building and Fire Research Laboratory, and the U.S. Department of Housing and Urban Development; Office of Policy Development and Research; Washington, DC 20410; September 2002, http://www.nist.gov/customcf/get_pdf.cfm?pub_id=860065 (accessed November 27, 2012).

[124]Partnership for Advancing Technology in Housing (PATH), a program of the U.S. Department of Housing and Urban Development.

(2) shell, (3) interiors, (4) plumbing, (5) HVAC, (6) fire protection systems, (7) electrical network, (8) communication and security networks, (9) fuel networks, (10) fittings, furnishings, and equipment. Level 1 elements are disaggregated into level 2 elements. For example, the level 1 element shell is disaggregated into three level 2 elements: (1) superstructure, (2) exterior enclosure, and (3) roofing. Where appropriate, level 2 elements are disaggregated into level 3 elements. For example, the level 2 element roofing is disaggregated into two level 3 elements: (1) roof covering and (2) skylights.

The service life data reported in Table 4-1 differ from the service life data reported in Tables 4-2 and 4-3. The reason for the difference is that the service life data reported in Table 4-1 are the "minimum" anticipated service lives for the building elements identified. The service life data reported in Tables 4-2 and 4-3 are "average" anticipated service lives for the building elements identified.

Building Element			Minimum Anticipated Service Life (Years)
Substructure	Foundation		20–60
Shell	Superstructure		20–60
	Exterior Enclosure	Grade Enclosure	20–40
		Walls	20–40
		Wall finish—easily renewable	4–10
		Wall finish—difficult to renew	20–60
		Doors	20–25
		Windows	20–25
		Door and window finish—easily renewable	3–10
		Joint sealants	3–10
	Roofing	Roof covering	12–25
		Skylights	15–25
Interiors	Vertical Space Dividers	Partitions	20–40
		Partition surface finishes—easily renewable	3–10
		Partition surface finishes—difficult to renew	15–25
		Doors	15–35
		Doors—easily renewable door paint	3–10
	Horizontal Space Dividers	Floor surfaces—bath and w.c.	4–20
		Floor surfaces—other	3–20
		Ceiling surfaces—private spaces; easily renewable	3–10
		Ceiling surfaces—private spaces; difficult to renew	15–25
	Stairs and Ramps	Stair surfaces—private spaces	3–15
Plumbing	Piping	Accessible	15–30
		Inaccessible	20–40
	Fixtures		15–25
	Fixture Trim		5–15
	Hot Water Heater		5–15
HVAC	Fuel Pump		10–15
	Pipes	Accessible	10–20
		Inaccessible	20–40
	Radiators		15–40

TABLE 4-1 Examples of Minimum Anticipated Service Lives of Typical Building Elements: ASTM Standard Guide E 2136

Building Element			Minimum Anticipated Service Life (Years)
HVAC	Fan Coil Units		15–20
	Pumps		10–15
	Boilers		10–20
	Mechanical Refrigerator	Machines	15–20
	Evaporative Coolers		8–15
	Furnaces		15–20
	Duct Work	Accessible	20–40
		Inaccessible	20–25
	Louvers		5–40
	Thermostats		5–15
	Exhaust Fans		5–15
Fire Protection Subsystems			5–30
Electrical Network	Conductors	Accessible	20–25
		Inaccessible	20–40
	Conduit		20–40
	Exposed Raceways		15–40
	Switches		5–15
	Sockets		5–15
	Overcurrent Protection	Circuit breaker	5–20
	Panels		20–40
	Switch Board		20–40
	Light Fixtures		5–15
Communication and Security Networks	Television Antenna		5–20
Fuel Network	Fuel Storage	Above ground	10–20
		Below ground	20–30
	Fuel Supply Piping	Accessible	20–30
		Inaccessible	20–40
Fittings, Furnishings and Equipment	Furnishings	Built-in furnishings	5–20
		Built-in furnishings surfaces—easily renewable	5–10

Table 4-1 Examples of Minimum Anticipated Service Lives of Typical Building Elements: ASTM Standard Guide E 2136 (*Continued*)

Building Element			Average Anticipated Service Life (Years)
Substructure	Foundation	Poured footings and foundations	200
		Concrete block	100
		Cement	50
Shell	Superstructure		Lifetime
	Exterior Enclosure	Basement floor system	Lifetime
		Framing (exterior walls)	Lifetime
		Exterior paint on wood, brick and aluminum	7–10
		Gutters and downspouts	30
		Siding—wood	10–100
		Siding—metal (steel)	50–Lifetime
		Siding—aluminum	20–50
		Siding—vinyl	50
		Exterior door (protected overhang)	80–100
		Exterior door (unprotected and exposed)	25–30
		Screen door	25–50
		Garage doors	20–50
		Window glazing	20
		Windows—wood casement	20–50
		Windows—aluminum casement	10–20
		Window screen	25–50
		Chimney	Lifetime
		Brick and stone walls	100+
	Roofing	Roof covering—asphalt and wood shingles and shakes	15–30
		Roof covering—tile	50
		Roof covering—slate	50–100
		Roof covering—sheet metal	20–50+
		Built-up roofing—asphalt	12–25
		Built-up roofing—coal and tar	12–30
		Asphalt composition shingle	15–30
		Asphalt overlag	25–35
Interiors	Vertical Space Dividers	Walls—drywall and plaster	30–70
		Walls—ceramic tile	Lifetime

Table 4-2 Examples of Average Anticipated Service Lives of Typical Building Elements: Housing Economics

Building Element			Average Anticipated Service Life (Years)
Interiors	Vertical Space Dividers	Framing (interior walls)	Lifetime
		Doors—not perfect hollow core	<30–Lifetime
		Doors—solid corewood	30–Lifetime
		Folding door	30–Lifetime
		Interior wall paint	5–10
		Interior trim and door	5–10
		Wallpaper	7
	Horizontal Space Dividers	Floor surfaces—vinyl sheet or vinyl tile	20–30
		Floor surfaces—oak or pine	Lifetime
		Floor surfaces—marble	Lifetime
		Floor surfaces—slate flagstone	Lifetime
		Floor surfaces—terrazzo	Lifetime
		Carpeting	11
		Ceiling suspension	Lifetime
		Acoustical ceiling	Lifetime
	Masonry	Fireplace and brick veneer	Lifetime
		Stucco and mantels	Lifetime
	Stairs and Ramps	Stairs	50–100
		Rails and disappearing stairs	30–40
Plumbing	Waste pipe	Concrete	50–100
		Cast iron	75–100
	Sinks	Enamel steel sinks	5–10
		Enamel cast iron sinks	25–30
		China sinks	25–30
	Faucets	Low quality	13–15
		High quality	15–20
	Water Heater	Gas	11–13
		Electric	14
HVAC	Air Conditioning Unit	Central unit	15
		Window unit	10
	Air Conditioning Compressor		15
	Humidifier		8

Table 4-2 Examples of Average Anticipated Service Lives of Typical Building Elements: Housing Economics (*Continued*)

Building Element			Average Anticipated Service Life (Years)
HVAC	Duct Work	Plastic	15
		Galvanized	30
	Rooftop Air Conditioners		15
	Boilers, Hot Water (Steam)		30
	Furnaces	Gas or oil fired	18
		Heat pumps	15
	Unit Heaters	Gas or electric	13
	Radiant Heaters	Electric	10
		Hot water or steam	25
	Air Terminals	Diffusers, grilles, and registers	27
	Induction and Fan-Coil Units		20
	Dampers		20
	Fans	Centrifugal	25
		Axial	20
		Ventilating roof—mounted	20
	Coils	DX, water, or steam	20
		Electric	15
	Heat Exchangers	Shell-and-tube	24
	Molded Insulation Molded		20
	Pumps, Sump and Well		10
	Burners		21
Fire Protection Subsystems	Smoke Detectors		12
	Smoke/Fire/Intrusion System		10
Electrical Network	Copper Wiring	Copper plated	100+
		Copper clad aluminum	100+
		Bare copper	100+
	Insulation	Armored cable (BX)	Lifetime
		Conduit	Lifetime

TABLE 4-2 Examples of Average Anticipated Service Lives of Typical Building Elements: Housing Economics (*Continued*)

Building Element			Average Anticipated Service Life (Years)
Shell	Exterior Enclosure	Solid core (painted) exterior door	40
		Refinish solid core (painted) exterior door	4
		Aluminum siding, 1st floor	80
		Refinish aluminum siding, 1st floor	5
	Roofing	Replace new over existing asphalt shingle roof	20
		Removal and replacement of asphalt shingle roof	40
Interiors	Vertical Space Dividers	Hollow core (painted) interior door	30
		Refinish hollow core (painted) interior door	4
		Fabric interior wall finish	10
	Horizontal Space Dividers	Vinyl tile flooring	18
		Acoustic tile ceiling	70
	Stairs and Ramps	Refinish masonry interior steps (painted)	6
Plumbing	Fixtures	Sink—stainless steel	40
		Valve set, sink, stainless steel	10
		Hot water storage tank	50
		New gasket, pipe and fittings, cast iron	50
		Pipe and fittings, cast iron	75
		Pump and motor assembly, sump pump	20
		Compressed air dryer	15
HVAC	Fuel Supply Systems	Fuel oil pump	15
	Heat Generation Systems	Gas house furnace	15
	Heat Rejection Systems	Chiller, reciprocal air cooled hermetic	20
	Heat Distribution Systems	Air handler, singlezone	15
	Heat Transfer	Air conditioner, indoor DX packaged, air cooled	20

Table 4-3 Examples of Average Anticipated Service Lives of Typical Building Elements: Whitestone

Building Element			Average Anticipated Service Life (Years)
Fire Protection Subsystems	Fire Protection Sprinkler Systems	Fire sprinkler head	20
Electrical Network	Electrical Service and Distribution	Transformer, dry, less than 15,000 V	30
	Lighting and Branch Wiring	Fluorescent lighting fixtures, 160 W	20
	Communication and Security Systems	Fire alarm control panel	15
	Special Electrical System	Generator, 1000 kW diesel	25

Table 4-3 Examples of Average Anticipated Service Lives of Typical Building Elements: Whitestone (*Continued*)

Building Element	Service Life in Years	Source
Plumbing		
Fixtures	15–25	ASTM E 2136
Enamel steel sinks	5–10	Housing Economics
Enamel cast iron sinks	25–30	Housing Economics
China sinks	25–30	Housing Economics
Stainless-steel sinks	40	Whitestone
Low-quality faucets	13–15	Housing Economics
High-quality faucets	15–20	Housing Economics
Hot water heater	5–15	ASTM E 2136
Gas water heater	11–13	Housing Economics
Electric water heater	14	Housing Economics
Hot water storage tank	50	Whitestone
HVAC		
Evaporative coolers	8–15	ASYTM E 2136
Central air conditioning unit	15	Housing Economics
Window air conditioning unit	10	Housing Economics
Air conditioner, indoor DX packaged, air cooled	20	Whitestone
Furnaces	15–20	ASTM E 2136
Gas or oil fired furnaces	18	Housing Economics
Gas house furnaces	15	Whistestone
Heat pumps	15	Housing Economics

Table 4-4 Summary of Baseline Measures of Service Life

Building Element	Service Life in Years	Source
Siding		
Wall finish—easily renewable	4–10	ASTM E 2136
Wall finish—difficult to renew	20–60	ASTM E 2136
Siding—wood	10–100	Housing Economics
Siding—metal (steel)	50–Lifetime	Housing Economics
Siding—aluminum	20–50	Housing Economics
Siding—vinyl	50	Housing Economics
Aluminum siding, 1st floor	80	Whitestone
Roofing		
Roof covering	12–25	ASTM E 2136
Roof covering—asphalt and wood shingles and shakes	15–30	Housing Economics
Roof covering—tile	50	Housing Economics
Roof covering—slate	50–100	Housing Economics
Roof covering—sheet metal	20–50+	Housing Economics
Replace new over existing asphalt shingle roof	20	Whitestone
Removal and replacement of asphalt shingle roof	40	Whitestone
Windows		
Windows	20–25	ASTM E 2136
Windows—wood casement	20–50	Housing Economics
Windows—aluminum casement	10–20	Housing Economics
Doors		
Doors	20–25	ASTM E 2136
Exterior doors (protected overhang)	80–100	Housing Economics
Exterior doors (unprotected and exposed)	25–30	Housing Economics
Solid core (painted) exterior doors	40	Whitestone

Table 4-4 Summary of Baseline Measures of Service Life (*Continued*)